COMPRESSION ALGORITHMS FOR REAL PROGRAMMERS

COMPRESSION ALGORITHMS FOR REAL PROGRAMMERS

Peter Wayner

ACADEMIC PRESS, a Harcourt Science and Technology Company

San Diego San Francisco New York Boston

London Sydney Tokyo

ACADEMIC PRESS
A Harcourt Science and Technology Company
525 B Street, Suite 1900, San Diego, CA 92101-4495, USA
http://www.apnet.com

ACADEMIC PRESS LIMITED
24-28 Oval Road, London NW1 7DX, UK
http://www.hbuk.co.uk/ao/

Morgan Kaufmann
A Harcourt Science and Technology Company
340 Pine Street, Sixth Floor
San Francisco, CA 94104-3205

Library of Congress Catalog Card Number: 99-65379
International Standard Book Number: 0-12-788774-1

Printed in the United States of America
99 00 01 02 CP 9 8 7 6 5 4 3 2 1

Contents

Preface ix

Book Notes xi

1 Introduction 1
- 1.1 Grading Compression Algorithms 6
- 1.2 Philosophical Hurdles 9
- 1.3 How to Use This Book 13

2 Statistical Basics 15
- 2.1 Huffman Encoding 16
- 2.2 Shannon-Fano Encoding 21
- 2.3 Entropy and Information Theory 22
- 2.4 Character Grouping Schemes 28
- 2.5 Conclusion 32

3 Dictionary Techniques 35
- 3.1 Basic Lempel-Ziv-Welch 36
- 3.2 Simple Windows with LZSS 40
- 3.3 Coding Notes 43
- 3.4 Variations 46
- 3.5 Commercially Available Standards 46
- 3.6 Conclusions 48

4 Arithmetic Compression 49
- 4.1 Three examples 51
- 4.2 Programming Arithmetic Coding 57
- 4.3 Products Using Arithmetic Coding 59

4.4 Conclusion 60

5 Adaptive Compression **61**
5.1 Escape Codes 62
5.2 Adaptive Huffman Coding 63
5.3 Windows of Data 65
5.4 Conclusion 66

6 Grammar Compression **67**
6.1 SEQUITUR 69
6.2 Code Compression 71
6.3 Conclusion 75

7 Programmatic Solutions **77**
7.1 PostScript 78
7.2 Conclusions 82

8 Quantization **85**
8.1 Basic Quantization 86
8.2 Adaptive Quantization 89
8.3 Vector Quantization 94
8.4 Dimension Reduction 98
8.5 Conclusion 100

9 Wavelet Transforms **101**
9.1 Basic Fourier Mathematics 103
9.2 Discrete Cosine Transform 106
9.3 Two-Dimensional Approaches 108
9.4 Other Wavelet Functions 111
9.5 Conclusion 123

10 JPEG **125**
10.1 JPEG Overview 126
10.2 Basic JPEG 127
10.3 JPEG Enhancements 133
10.4 Lossless JPEG 134
10.5 Progressive Transmission 136
10.6 Hierarchical Transmission 137
10.7 Conclusions 139

11 Video Compression **141**
11.1 Pixel Details . . . 142
11.2 Motion Estimation . . . 144
11.3 Quantization and Bit Packing . . . 150
11.4 MPEG-2 . . . 153
11.5 Conclusions . . . 153

12 Audio Compression **157**
12.1 Digitization . . . 159
12.2 Subband Coding . . . 160
12.3 Speech Compression . . . 161
12.4 MPEG and MP3 . . . 161
12.5 Conclusion . . . 164

13 Fractal Compression **165**
13.1 Conclusion . . . 167

14 Steganography **171**
14.1 Statistical Coding . . . 172
14.2 JPEG and JSteg . . . 173
14.3 Quantization . . . 174
14.4 Grammars . . . 175
14.5 Conclusions . . . 177

A Patents **179**
A.1 Statistical Patents . . . 180
A.2 Dictionary Algorithm Patents . . . 183
A.3 Arithmetic Algorithm Patents . . . 194
A.4 Adaptive Algorithm Patents . . . 209
A.5 Grammar Algorithm Patents . . . 213
A.6 Quantization Algorithm Patents . . . 214
A.7 Image Algorithm Patents . . . 215
A.8 Fractal Algorithm Patents . . . 217
A.9 Other Patents . . . 219

B Bibliography **221**

Index **235**

Preface

The team of people at Morgan Kaufmann have been incredibly gracious with their time and encouragement. I'm glad for all of their support through this manuscript. They are: Tom Stone, Thomas Park, Chuck Glaser, Shawn Girsberger, Paige Whittaker, and Gabrielle Billeter. Carsten Hansen provided a technical review of the book and contributed many important and significant improvements to the book.

There were others who helped in the world beyond the text. The staff at Tax Analysts were kind enough to coordinate my consulting schedule with the demands of putting out a book. Anyone would be lucky to work for a company that was so understanding. Also, friends like August and Lise Uribe were kind enough to let me stay at their house while I was writing a large portion of this book.

Finally, I want to thank everyone in my family for everything they've given through all of my life.

Peter Wayner
Baltimore, MD
July 1999
`pcw@flyzone.com`
`http://access.digex.net:/~pcw/pcwpage.html`

Book Notes

The copy for this book was typeset using the LaTeX typesetting software. Several important breaks were made with standard conventions in order to remove some ambiguities. The period mark is normally included inside quotation marks like this, "That's my answer. No. Period." This can cause ambiguities when computer terms are included in quotation marks because computers often use periods to convey some meaning. For this reason, my electronic mail address is "pcw@access.digex.com". If the period comes outside of these quotation marks, it was done to prevent ambiguity.

Hyphens also cause problems when they're used for different tasks. LISP programmers often use hyphens to join words together into a single name like this: `Do-Not-Call-This-Procedure`. *Unfortunately, this causes grief when these longer words occur at the end of a line. In these cases, there will be an extra hyphen included to specify that there was an original hyphen in the word. This isn't* hyper-compatible *with the standard rules that don't include the extra hyphen. But these rules are for readers who know that* self-help *is a word that should be hyphenated. No one knows what to think about* `A-Much-Too--Long-Procedure-That-Should-Be-Shortened-For-Everyone.`

Finally, I hope that all readers who see problems with the book will take the time to write me. Any errors or omissions will be reported on an errata sheet.

Chapter 1

Introduction

The science of compressing data is the art of creating shorthand representations for the data— that is, automatically finding abbreviations; i.e. yadda yadda yadda, etc.

All of the algorithms can be described with a simple phrase: Look for repetition, and replace the repetition with a shorter representation. This repetition is usually fairly easy to find. The letters "rep" are repeated eight times in this paragraph alone. If they were replaced with, say, the asterix character (*), then two characters would be saved eight times. It's not much, but it's a start.

The algorithms succeed when they have a good model for the underlying data. They can even fail when the model does a bad job of matching the data. The model of looking for three letters like "rep" works well in some sentences, but it fails in others. The art of designing the algorithm is really the art of finding a good model of the data that can also be fit to the data efficiently.

The algorithms in this book are different attempts to find a good, automatic way of identifying repetitive patterns and removing them from a file. Some work well on text data, while others are tuned to images or audio files. All of them, however, are far from perfect. If an algorithm has a strength, then it will also have a weakness. The best algorithm for some data is often the worst for other types of data. To paraphrase Abraham Lincoln: You can compress all of the types of files some of the time and some of the types of files all of the time, but you can't compress all of the types of files all of the time.

The basic ideas in this book can be broken into these groups:

Statistical Approaches The program analyzes the file and looks for the most common bytes, characters, words, or other items in the file. The most common items are given short abbreviations, while the least common ones are given longer abbreviations, often longer than the original item. When all of the replacing is finished, the final file is smaller because the common items get shorter representations.

These solutions usually work fairly well with text and other data that break apart into consistent parts. Text files are made up of characters and words, so there are natural ways to compute the statistics.

The statistical approaches are also used as frosting in more complicated compression algorithms like the JPEG method for reducing photographs. In this case, the first part of the algorithm converts the image into a collection of coefficients, and then these coefficients are packed with a statistical method that takes advantage of how often they occur.

Dictionary Approaches This solution builds a list of the top n words, phrases, or chunks of data in a file and then replaces these with the numbers between 0 and $n-1$. The value of n is usually a power of 2, such as 2^8, because this allows the data to be packed completely in binary form. Storing the numbers between 0 and $2^n - 1$ takes n bits.

These solutions also work especially well with text where there are many common phrases that repeat. Images are often difficult to handle because there are fewer common patterns and there are often subtle differences between their different appearances. A picture of a flock of 10,000 seagulls may have 10,000 similar objects, but they are often different in subtle ways. Not only are the birds different creatures, but they may be posed in slightly different ways and be illuminated in different forms. Words are exact, but images aren't.

In some cases, the image file compression programs use dictionary approaches after they've converted the images into an intermediate format. The dictionary approach is used for the final round.

Pattern-Finding Approaches Some data come with regular patterns that can be extracted by simple programs. For instance, fax machines send simple black-and-white images. These are easily compressed with a solution known as run length encoding, which counts the number of times a black or white pixel is repeated. The best way to illustrate this is with a quick example. Imagine that 'BBBBBWWWWWWWWBBBBWBWWWWWWWW' is a part of an image, where 'B' represents a black pixel and 'W' represents a white pixel. This is compressed to be 0574117, with the first zero implying that there are no white pixels at the beginning.

Run length encoding is the most common example of solutions based upon identifying patterns. In most cases, the patterns are just too complicated for a computer to find regularly.

Smart Packing Solutions Is there a difference between the intelligent use of space and compression? In some cases, no. Programmers have long known that it is silly to use double precision values or long integers when regular floating point numbers or integers will serve the purpose. Waste not, want not.

This careful analysis can be surprisingly useful. David Kopf takes this one step further with an algorithm that carefully adjusts the number of bits used for each value. It starts with say 16 bits for the first number in a sequence. If this too many bits, then it reduces the size by i bits for the next value. If it is too few, then it increases the size by j bits and repeats the value. In some cases, the algorithm uses a fixed i and j, and, in others, it uses a preset number.

Approximate Solutions Some compression routines save a great deal of space by creating approximations to the data. When a file is uncompressed, it is usually close to the original file but not exactly the same. These solutions are commonly used to compress images and audio files because both of these are usually just digital approximations already. Another level of approximation is often perfectly acceptable to the user who often might not even notice the difference.

Compression routines that always reproduce the same data when a file is decompressed are known as lossless. *Those that only return a close approximation are called* lossy.

In some cases, the result of decompressing a file that was compressed with a lossy algorithm can be better than the original. Lossy algorithms often succeed when they avoid storing all of the details about the noise in an image or a sound recording. The user often wants to get rid of this extraneous detail and a lossy or approximate algorithm can often accomplish this as a side effect.

In other case, the approximation can introduce unpleasant effects. The JPEG algorithm (described in Chapter 10) can do a great job with many photographs, but it can introduce blocky, gridlike effects when it is pushed to generate the smallest possible files. These occur because the algorithm tries its best to approximate each 8×8 block of pixels.

These algorithms are never used with many types of data because an approximation could be fatal. Bank records and other financial information are a good example of information that should be reproduced exactly. Medical data like CT scans or digital X rays are also often reproduced exactly because doctors worry about their patients.

Adaptive Solutions Some programs try to adapt to local conditions in a file. For instance, a novel will often include bit characters in certain chapters. A good, adaptive dictionary approach will include this character's name for this chapter but replace it with something more common in the other chapters when the other chapters are compressed.

There are a vast number of adaptive schemes, and each is tuned to particular type of data. Some solutions are more general than others, but all have strengths and weaknesses. Most solutions can be made adaptive by focusing them on some moving window of data. For instance, the statistical approaches can adapt by looking at the text in smaller blocks of n bytes. These windows can be either fixed or sliding.

Programmatic Solutions One of the most important solutions is to create an entirely new programming language for encoding your data. The PostScript language for printing information is one of the best examples on the market. It takes often large images and represents them in a simple way that saves huge amounts of space. A black line might take up thousands of black pixels, but PostScript describes it with its endpoints. A single letter may contain hundreds of pixels, but PostScript describes it with one character.

Many of the multimedia formats like Macromedia's Flash also include a fair amount of programmatic structure. This allows them to represent a complicated animation in substantially less space than a movie would take. Instead of encoding each frame as a collection of bits, the screen is described as a collection of visual objects that move and interact. This can take substantially less space.

These solutions can be very efficient, but they do not work with arbitrary files. A programmer must create a language for a particular type of file, and this usually can't be done automatically. A bigger problem is the fact that there's often no easy way to compress a file unless it is created in the special format. For instance, a word processor can create a PostScript file by converting the text into PostScript instructions to draw particular characters at particular locations. But it is difficult to take a fax image and convert it into such a PostScript file. That requires analyzing clumps of pixels and determining which ones form which characters. This may be possible, but it isn't easy.

The good news is that these programmatic solutions can save plenty of space. The bad news is that they require plenty of planning. Each of the other algorithms described above will go to work automatically on any file and that just won't work in this case. The programmatic solutions also need much more computational power. The earliest PostScript printers had more computational horsepower than the computers they served. Multimedia formats may take up much less space than a simple movie, but they must be rendered on the fly. The computer must compute the position and look of each object for each frame. This means that the objects and their interaction can't be too complicated or the animation will be too slow.

Of course, programmatic solutions can also fail, like all compression algorithms. Simple images or texts can take much more space when they are converted into a richer language like PostScript. In a sense, ASCII text is a great compression language that most line printers instinctively know how to render on the page. They just keep spitting out the letters until a carriage return sends them to the next line.

Wavelet Solutions The term *wavelet* includes many different types of repeated patterns. A wavelet-based compression scheme will try to approximate the data by adding together a number of basic patterns known as wavelets that are weighted by different coefficients. If the program does a good job, then the sum is pretty close if not exactly the same as the original. Then the coefficients can take the place of the original and save plenty of space.

While the term wavelet is pretty generic, the most common wavelet in use is the cosine function. This repeating transcendental function should be fairly familar to most people who have completed much of high school trigonometry, because it repeats ad infinitum. Newer wavelets have more complicated shapes, and they are often set to zero outside of a small region of interest.

Many engineers are also familar with Fourier series, which represent a function as a weighted sum of cosine functions with various speeds of repetition. The wavelet compression schemes are all extensions and generalizations of these basic ideas discovered well in the past. The main difference is that the original solutions used an infinite number of terms in order to make a deeper mathematical point. These solutions are engineered to do as good a job as possible with a minimal number of coefficients and waveforms.

Much of the recent work on wavelets has also explored how to localize the different wave forms. The basic Fourier approach used functions defined over the entire real number line. New, discrete versions focussed on only one bounded region. The best new algorithms subdivide the region and use one set of wavelets for one region and one set for another. In a sense, the wave functions are only defined over these subregions, and this leads to the term *wavelet*. These decompositions of the region can be quite complicated, and they are largely responsible for the success of the latest algorithms.

Of course, the biggest class of compression algorithms is a hybrid of some of the types described above. A common solution, for instance, is to build a dictionary of common phrases and then use statistical analysis to determine which are more common than the others. Some of the image approximation algorithms attempt to match preset patterns to the image.

Many software solutions often include several algorithms inside of them and test each algorithm against a file. The algorithm with the best performance is the one that is finally used. This sort of adaption is common with commercial packages.

1.1 Grading Compression Algorithms

The easiest way to measure a compression algorithm is to feed it a file and see if it can make it much smaller. The amount of shrinkage may be represented as a percentage ("The file is 32% of the original.") or as a factor ("The compression algorithm reduced it 5 to 1."). This sort of absolute measure is the best test.

Saying anything more general about a compression algorithm is difficult and fraught with caveats. One algorithm may produce great compression on one file but fail to compress another at all. In many cases, it's hard to produce a good explanation. Some algorithms have a good model that provides solid estimates of their performance while others are less consistent. It's difficult to go beyond generalities like "This algorithm is good for most images."

While it may not be possible to provide good, general guarantees for each algorithm, it is often possible to analyze the performance of algorithms in different terms. Here are some of the other criteria that people use:

Overhead Each algorithm will usually attach a certain amount of extra data that it needs to decompress a file. For instance, the statistical approaches append a list of the common characters, words, or phrases to the file. The dictionary solutions must arrange for the same dictionary to be available when a file is uncompressed.

This overhead is often much smaller than the amount of space that is saved, and there is no problem when it is added. But there are many times when the overhead is larger than the amount being saved, and the size of the file grows.

In some cases, the overhead also includes a small decompression program that turns the entire file into what some call a *self-extracting archive*.

Variation Some algorithms give consistent performance. Others do very well on some types of files. The amount of variation is rarely codified or easily predicted, but it is possible to come up with empirical results based upon past performance.

Degradation Many of the approximation algorithms introduce some degradation in the data in the compression process. The amount of this degradation is often controllable, and the better algorithms offer a tradeoff between degradation and file size. Smaller files mean more degradation and vice versa.

In many cases, the degradation can also be characterized and maybe even measured. The JPEG image compression algorithm wlll tend to replace areas of similar colors with large blocks.

Patent Status Some compression algorithms are patented. Some of these patents have some legal strength. Others are weakened by challenges to their applicability. The legal status of many of these algorithms has tripped up others. Microsoft, for instance, lost a large lawsuit filed by a small company that contended that Microsoft infringed on its patents. The best solution for many developers is to license a library of compression routines, because the authors of the compression routines have already done much of the patent evaluation. They should be paying the patent holders for you.

Compression Speed There are wide differences in this value. Some programs spend a fairly large amount of time creating complicated statistical profiles of the data. Others make a quick pass over the data. If the algorithm is going to remain competitive, then the extra compression time must result in smaller files.

In many cases, developers don't care about compression speed, but they do care about decompression speed. Movies, photos, and audio files are usually compressed once by the artist and then decompressed many times by the different viewers. Developers don't care about spending a few extra hours or even days, if the result is substantially smaller.

In others cases, compression time is important. Some programs will record audio or video files directly to a computer disk. Any compression helps save space and makes the software more usable. But this compression can't take too much time because the software must work in real time and continue to record incoming data.

Decompression Speed Naturally, decompression speed matters to someone decompressing a file. This is especially important today because many companies are looking for ways to distribute movies and sounds over digital delivery systems. Small files are great, but the computer should be able to decompress them on the fly. If the decompression process is too slow, then the uncompressed file must be stored on a hard disk before it can be played. In many cases, there isn't enough space on the disk to do the job effectively.

Many computers come with special processors designed to decompress files compressed with the MPEG video algorithm. This special hardware is necessary to display the decompressed images in real time.

Online Suitability Some of the compression algorithms were designed to compress data flowing over the Internet. While all compression algorithms that work for data on the hard disk will also help keep transmission costs down, some algorithms offer better performance when the Internet fails to deliver parts of the data stream.

For instance, the RealAudio format is designed to recover smoothly if a packet of data from the middle of the file disappears along the route. Other basic sound files fill each packet with all of the information about n seconds of the file. This guarantees that there will be n seconds of dead air if the packet doesn't arrive. RealAudio, on the other hand, interleaves the information from k different blocks and packs it into successive packets to ensure that no packet contains more than $\frac{n}{k}$ seconds. k is set to be large enough to ensure that many glitches can be smoothed over.

Clearly this solution requires a bit of preprocessing. While the compression process might not be too complicated, it does take time to start up. If information from k different blocks is interleaved, then k packets, must be processed before the sound can emerge. This means that replay is delayed by k packets, and this is often known as the *latency* of the algorithm. This isn't a great loss if a user is replaying an old recording, but it makes the solution unsuitable for use on a telephone. The delays make it impossible to hold a decent conversation.

1.2 Philosophical Hurdles

Compression is not an easy subject for people to master. Although many of the algorithms in this book are very straightforward, there are larger issues about compression that are not well defined. This often means that it is easy to grasp how an algorithm will work, but it is hard to understand why it must fail. Understanding compression algorithms at a larger level means grappling with some fairly philosophical distinctions about information and knowledge. While they're not easy nor essential, they are the only way to gain true mastery of the subject.

One of the first hurdles for many programmers is understanding that compression is possible at all. Today, many people learn of compression before learning to program because the practice is so common. Most people install software today with popular decompression packages. In the past, though, many learned to program before learning about compression.

My first instinct was that compression was impossible. If data were being removed, then information was being lost. Clearly, the amount reconstructed at the other end when decompression took place couldn't be as rich or robust as what went into the compression routine. Information had been lost. How was it possible to compress a file containing the name, rank, and serial number of a soldier? Sure, the name could be compressed by replacing the first name with an initial, but then it would be impossible to figure out what the first name was later.

This instinct is both correct and incorrect. It is true that general, all-purpose compression is indeed impossible. There is no algorithm that will make all files smaller. There's no algorithm that can guarantee to shave even one bit off all files. This is not hard to prove in an informal way. Let T_n be the set of all files with exactly n bits. There are 2^n possible files in T_n and $2^n - 1$ possible files in the aggregation of all T_k where $k < n$. If there were an algorithm that compressed all of T_n into the smaller files, then at least two files would end up being compressed into the same smaller file. How could a decompression routine tell which was which when it decompressed them? How will the decompression routine tell a file compressed from one T_n from an uncompressed version from a T_k? Remember, it isn't possible to add extra bits to indicate the original size of the file or whether it is compressed because this overhead will make the files larger. At least half of T_n is being compressed by only one bit in this scheme, and adding even one bit of overhead will alleviate any gain from this procedure.

This is an informal proof, and it has some nasty implications. The worst is that half of all files can't be compressed by more than one bit. Three-quarters of the files can't be compressed by more than two bits. Seven-eighths can't be compressed by more than three. That's pretty depressing news.

The good news is that while the compression is difficult if not impossible in the general case, it is usually quite useful for most of the things we do all day. Most of the files, for instance, are filled with such rich patterns and repetition that the simplest program can identify them and use them to shrink the file. Text is filled with endless streams of only a few words. Images have huge chunks devoted to the same or similar colors.

This means that it's possible to design compression algorithms that do little with the text files with strings like "jaeujxm kjsaA9kk3*" and do a lot with strings like "Bobbie bobbed her hair and bobbed for apples." There are many different ways that words can be combined to form sentences, paragraphs, and books, but there are many, many, many more ways that the letters can be combined in random ways. It's not hard to exclude these cases.

This lack of guarantees may be hard for some people to grasp. Most users are accustomed to using popular programs like PkZip or Stuffit that regularily squeeze files down to half their size. Compressing most common files is fairly easy because most data are largely repetitive and filled with plenty of opportunities to find some shorthand representation. Image files are often quite large, and there is usually plenty of opportunity to find a shorthand approximation for parts of the file.

Another common mistake that many people make is assuming that files can be continually compressed. For instance, if some algorithm cuts the size of files in half, then what happens if it is applied again to the result? Will the resulting file be one-quarter of the size? Probably not. A good compression function should do all of the work in the first pass. This isn't always the case, but it should be.

Why can't a file be compressed and recompressed until it is very small and manageable? Here are two explanations. The first extends the informal proof from above and makes the same basic argument: If all files could be compressed to some size like n bits, then there could be at most 2^n files out there. This is still a fairly large number of even moderate sized n, but some are going to start bumping into each other in the compression process and end up with the same compressed version.

Chapter 2 contains a long description of Claude Shannon's theory of information. This gives a more theoretically sound definition of why compression can't go on forever.

The second reason is more difficult to explain. Compression stops because the algorithm stops finding patterns to exploit. After each pass, the file should grow more and more random and less and less filled with patterns. While there will be counterexamples of a few pathological examples, most files will stop being compressable after the randomness peaks.

Here's a quick example drawn from Morse Code. My first name in Morse Code is, "Dit Dah Dah Dit pause Dit pause Dah pause Dit pause Dit Dah Dit". Compressing this is easy because there are only three words in the string. If they're replaced by the integers, 0, 1 and 2, then the result is 12210102010121. Now, the three letter patterns of "Dit" and "Dah" are not repeated any more. Deeper patterns, however, might still be found.

A Counter example

The last part tried to explain how there was no free lunch when compressing data. A fairly straightforward proof showed that no function could work all of the time. Still, there are many philosophical connundrums in the area. To add some confusion, this section will describe a function that will compress every file into no more than, say, 512 bits. The catch is that the section won't actually describe how to build such a function.

The function's actions, at least in the most abstract sense, are pretty simple. It examines some file and determines who created it. Then it figures out the exact time and place of its creation. The time can be stored by using 128 bits to encode the number of seconds since the creation of the universe. Yes, this has something similar to a Year 2000 problem, but if you care deeply about this you can use 256 bits. Another 128 bits for the lattitude and longitude should be able to pinpoint the computer chip creating the data. 32 bits should encode the altitude to sufficient precision. If necessary, the person creating the data can be encoded with about 64 bits. This leaves about 32 bits for spares.

Some say such a compression function exists in the tape vaults of the government agencies with the job of spying on the world. Their vast collection equiptment continually gathers phone calls, e-mail, and other data before tagging each bit with a timestamp.

This magical function does a good job with all files created on Earth and ignores the files that never existed and thus never really matter. It treats the world as if it were filled with monkeys pecking away at typewriters. The compression code just picks the right monkey at the right time to act as a surrogate for the file.

This magical function doesn't violate the precept from the previous section proving that no function can compress all files, because the function won't compress all files. It will only work on those that get created on Earth by humans. It might not have much of a market on other solar systems, but it should sell well here on Earth.

The rub is that there's no easy way to build such a function. There's not really any known hard way either. Such a function would need to have the ability to reconstruct the actions of any particular human at any particular time. Humans are pretty predictable and repeat themselves again and again and again, but there are limits to what even the most clever computer program can do.

An interesting question might be: How big is such a program for compression and decompression? It could easily be pretty large. Each person would need to have a copy of the program around to decompress arbitrary files. If the program were bigger than the average size of the files used by most people, then it would be a waste of space. Each file might be compressable down to 512 bits, but the program would be so large that it would wipe out any space saved by the process.

This example is an extreme version of most compression programs. Designers of regular algorithms face the same tradeoffs in creating programs for everyday use. Extra intelligence for compressing functions adds both additional time in compression and size to the decompression routine. Either may not be worth the extra trouble.

Kolmogorov Complexity

One of the more theoretical ways to measure the "complexity" of a file is to compute the amount of space it takes to hold a computer program that would recreate the file. This is the same thing as asking what is the final size of a self-extracting archive created by the best compression function. This measurement is often used in some theoretical proofs about the limits of what computers can do.

The definition itself doesn't offer many lessons on how to build compression functions that actually work, but it does provide some inspiration. While all of the functions described in this book are self-contained, well defined, and pretty practical, there is no reason why there can't be more programs that do a better job.

For instance, imagine the file containing all 65,536 possible 16-bit numbers in sorted order. There is no statistical pattern that would be easily extracted by the algorithms in Chapter 2. Each number only occurs once, and all of the digits are used with close to the same frequency. Each 7 is just as likely to be followed by a 3 as a 5. In binary form, all of the bits occur equally often in each of the positions. None of the dictionary algorithms from Chapter 3 would do much good. The numbers aren't repeated. But a program written with one loop could duplicate the entire list.

The Kolmogorov complexity of a file is rarely used in practical settings. The number of different programs that could generate a particular output are so great that it is hard to check them all and determine which was the smallest. It also makes it difficult to write a program that will automatically find another program to simulate the data. In fact, it is really intended to be a theoretical construct.

But the Kolmogorov complexity can serve as an inspiration to programmers. It should hold out the possiblity that more and more complicated descriptions and software packages should be able to extract more patterns from the data. Reducing these patterns should add more compression.

1.3 How to Use This Book

This book covers most of the generic ideas for compressing data. Most compression on the Net uses algorithms covered in the first several chapters on statistical approaches (Chapter 2) and dictionary approaches (Chapter 3). These are the foundation of the field and the best place to begin.

Image compression is dominated by wavelet methods or, to be more exact, the discrete cosine transform. This is explained in theory (Chapter 9) and in detail in Chapters 10 and 11. Chapter 9 on wavelets describes the theory of how the wavelets provide some basic structures that can model all images, and Chapters 10 and 11 describe how the bits are actually packed.

Audio compression is also becoming increasingly important these days. The Internet makes it possible for people to exchange files. The basic methods for compressing audio files are described in Chapter 12. In some cases, they use techniques like the discrete cosine transform or the wavelet transform described in Chapter 9.

The rest of the book helps fill in details that are important in other cases.

Chapter 2

Statistical Basics

The most basic solutions for compressing files revolve around computing how often particular patterns can be found in the file and then replacing the most common patterns with shorter patterns and replacing the least common ones with longer patterns. This shrinks the file because more is saved by replacing the common patterns with shorter ones than is lost by replacing the least common patterns with longer ones. In the end, the tradeoff is worth it.

The solutions described in this chapter are particularly useful for compressing files filled with English text or other samples where there is a wide range of usage. The letter 'e' is quite common in English, while the letter 'z' is relatively rare. This provides plenty of opportunity to save space by replacing the 'e' with a shorter string of bits.

The statistical techniques here are also used to save more space in image compression where they are combined with wave-based solutions. See Chapters 10 and 11.

These solutions are usually described as applying to letters or bytes. The statistics can be gathered for any collection of characters, bytes, tokens, or patterns of bits. The biggest problem is that both the compression and decompression routines must have access to the same list of tokens or bits. If ASCII characters or bytes are used, then it is easy to arrange for the decompression routine to have access to the same list. If arbitrary tokens, words, or patterns are used, then a separate table must be shipped to the decompression routine. If the same table is used frequently, then the overhead is minor because the table can be stored with the decompression routine. If a new table is computed for each file, then it will add significantly to the overhead involved with the file. Of course, the effect of the overhead must be compared against the size of the savings. A compression algorithm that needs 10,000 bytes of overhead, but chops a 1 million byte file in half, is still saving 490,000 bytes.

Strategies for sophisticated grouping of characters begin on page 28.

For now, the best way to begin is by assuming that each file is made up of ASCII characters or bytes, and compression will occur by acting on these characters. The most common letters will be replaced by shorter bit strings, which may be as short as one bit long but will probably be about four or five bits long. The least common letters will get longer replacement bit strings that may be as long as 12 bits. The number of bits depends upon the probability distribution of the characters in each file.

For notational simplicity, let A stand for the "alphabet" or collection of characters in the file. Let $\rho(x)$ stand for the probability with which the character x is found in a file. This is normally represented as the number of times x is found divided by the total number of characters in the file. It may be weighted, however, by other considerations like the number of bits used to represent a "character" or symbol. Most English words are stored as ASCII text, which uses the same number of bits per letter, but other languages may use a flexible amount. Also, it is possible to use these statistical methods on words instead of characters.

The first three sections of this chapter describe three different ways to achieve the same end goal. The algorithms are all very similar and produce close to the same results with files. All are extensions of the work that Claude Shannon did on characterizing information and developing a way to measure it. This work is described later in the chapter to provide a theoretical basis for understanding the algorithms, how they work, when they will succeed, and when they will fail. The final section provides some strategies for optimizing the algorithms with data.

Information theory starts on page 22.

2.1 Huffman Encoding

The first approach is commonly known as "Huffman encoding" and named after David Huffman [Huf52]. It provides a simple way of producing a set of replacement bits. The algorithm is easy to describe, simple to code, and comes with a proof that it is optimal, at least in some sense.

A tree *is a data structure produced by linking* nodes *with* edges. *The top of the tree is the* root. *Some* interior nodes *have* descendants. *Those with none are* leaves.

The algorithm finds the strings of bits for each letter by creating a tree and using the tree to read off the codes. The common letters end up near the top of the tree, while the least common letters end up near the bottom. The paths from the root of the tree to the node containing the character are used to compute the bit patterns.

Here's the basic algorithm for creating the tree. Figure 2.1 shows how it works for a simple sample.

1. For each x in A, create a node n_x.

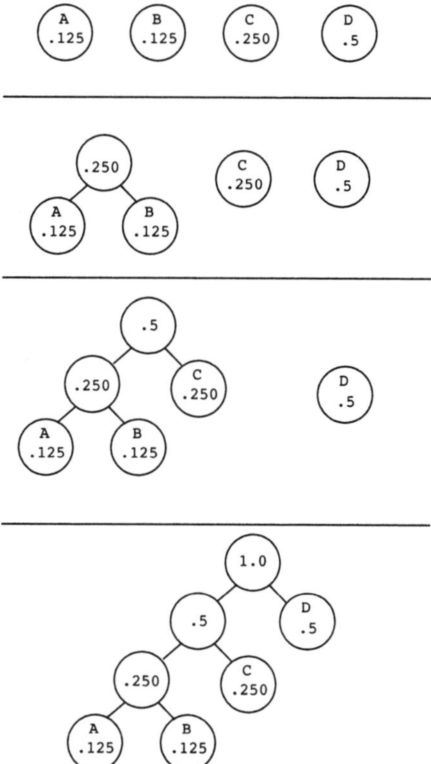

Figure 2.1: This illustrates the basic algorithm for creating a Huffman tree used to find Huffman encodings. The four levels show the collection of trees in T after each pass through the loop.

2. Attach a value $v(n_x) = \rho(x)$ to this node.
3. Add each of these raw nodes to a set T that is a collection of trees. At this point, there is one single node tree for each character in A.
4. Repeat these steps until there is only one tree left in T.
   (a) Check the values $v(n_i)$ of the root node of each tree. Find the two trees with the smallest values.
   (b) Remove the two trees from T. Let n_a and n_b stand for the two nodes at the roots of these trees.
   (c) Create a new node, n_c, and attach n_a and n_b as the left and right descendants. This creates a new tree from the old two trees.
   (d) Set $v(n_c) = v(n_a) + v(n_b)$.
   (e) Insert the new tree into T. The number of trees in T should have decreased by one for each pass through the loop.

The final tree is used to find bit patterns for each letter by tracing the path from the root to the leaf containing that node. The left path coming off each interior node is designated a '1' and the right path is designated a '0'. The string of values for each branch along the path will be the value used to replace the character.

Figure 2.2 shows a small tree with four characters that was constructed in Figure 2.1. The addresses for the paths are shown here. The letter 'A' has a path '111' that goes from the root to the leaf node. 'B' has a path with addresses '110'; 'C' has a path '10'; and 'D' has a path '0'.

Encoding is straight-forward. The word 'DAD' becomes '01110', the word 'CAB' becomes '10111110', the word 'BAD' becomes '1101110', and the word 'ADD' becomes '11100'. Notice how the words with the common letter 'D' end up shorter.

Decoding is also straightforward. Start with the first bit, and walk down the tree using the other bits to steer your path. Eventually, you'll get to a leaf node. That's the first character. Repeat this process until it's all decoded.

The strings produced by this method have several important characteristics:

Length The most common characters are the last ones chosen by the tree-building process. This means that the most common characters are closer to the top and have shorter paths from root to leaf. That means their replacement codes are shorter, which will be the basis of compression.

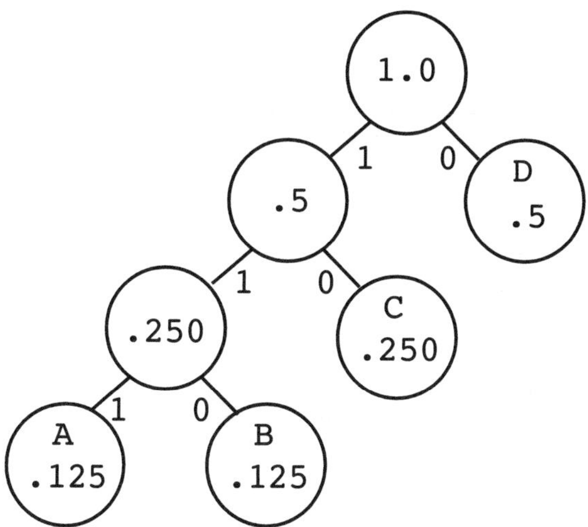

Figure 2.2: The final tree from Figure 2.1 with addresses given to the nodes. The left descendent of a node is the '1' and the right is the '0'. The code for 'A' is '111', 'B' is '110', 'C' is '10', and 'D' is '0'.

Prefixes The codes in this system have an interesting and essential feature. No code for one character is a prefix of another. This is essential to decompressing a file by matching up the strings of zeros and ones with a character. These strings are a variable length, so it is not possible to use the length to tell when a code ends. The commonly used ASCII code, for instance, encodes characters as eight bits. The end of each codeword is easy to find. The variable length of Huffman codes, however, makes it impossible to use this trick. The fact that no code is a prefix of another means that no impossible decisions must be made when decoding a string. For instance, imagine that 'A' was encoded as '111' and 'E' as '11', then it would be impossible to tell whether '11111' stood for 'AE' or 'EA'.

Optimal The Huffman codes constructed from such a tree are considered optimal in the sense that no other set of variable-length codes will produce better compression. The most common characters will be closer to the top, so they will have the shortest strings. If two characters from different levels are swapped, then the less common character will get a shorter code at the expense of a more common character getting a longer code. This isn't the right recipe for shrinking a file.

Another way to understand why the tree is optimal is to consider the construction process. At each step in the loop, two trees are taken from T and combined into one. This step effectively adds one bit to the code of every character that was already aggregated into these two trees. Which trees are chosen to have their codes extended by one bit? The two with the least common characters combined.

Some people use a stricter definition for "optimal". They define this to mean a code is optimal if it reduces the stream to a perfectly random string of bits in which a '1' is just as likely to occur as a '0'. In this case, the Huffman codes are optimal only if the probabilities that characters will occur are all powers of two. This is the case in Figure 2.2. 'A' occurs with probability 2^{-3}. If the probabilities are not powers of two, then the compression produced by this tree will still be the best possible with variable-length codes, but it won't produce a compressed file with perfect randomness.

Perfect randomness is often described as the entropy *of a data source. Page 22 discusses this term and the mathematical model underneath it.*

Fragile The variable length codes are a great way to squeeze space out of a file, but they are also fragile. If one bit disappears from the file or is changed, then the entire file after the disappearing bit could be corrupted. The only solution is to break the file up into smaller subfiles and keep track of the start of each subfile. Any corruption will be limited to a subfile.

Fast coding

The simplest way to encode data with this method is to create a table with two values for each character. One value would be the length of the code in bits, and the second would be the code itself stored as an integer. Here's what the codes from Figure 2.2 might look like. The codes are stored as eight-bit bytes. In practice, 32-bit-long integers might make more sense for modern processors.

Letter	length	code
A	3	11100000
B	3	11000000
C	2	10000000
D	1	00000000

The code for compressing the files should use the logical OR operation to insert the code and the left and right shift operations to line them up. This compression is usually done with values the size of the processing chip's registers in order to make it operate as fast as possible.

Here's some pseudocode:

```
integer outputWord=0;
```

```
integer bitsUsedInOutput =0;
integer maxBitsInWord=32;
integer[] codeWords; //Where the codes are found.
integer[] codeLengths; // How long they are.
integer temp;
char codeMe;
while more characters are uncoded do begin
     codeMe=get next character;
     temp=codeWords[codeMe];
     if (bitsUsedInOutput + codeLengths[codeMe]
         > maxBitsInWord)  then
      begin
        temp=rightshift(temp,bitsUsedInOutput);
        outputWord=OR(outputWord,temp);
        store outputWord in file;
        outputWord=leftshift(codeWords[codeMe],
            maxBitsInWord- bitsUsedInOutput);
        bitsUsedInOutput
          =maxBitsInWord-codeLength[codeMe]+bitsUsedInOutput;
      end else begin
          temp=rightshift(temp,bitsUsedInOutput);
          outputWord=OR(outputWord,temp);
          bitsUsedInOutput=bitsUsedInOutput
            + codeLength[codeMe];
      end;
   end;
```

Decoding can be a more complicated process. The simplest approach is to keep the tree in memory and walk down the tree pulling bits off as you go. This process can be slow if it is done bit by bit. The process can be accelerated by aggregating the lookup for the first several bits. A table created with 2^n entries for the first n bits will speed things up.

2.2 Shannon-Fano Encoding

The Shannon-Fano compression algorithm is a close cousin to the Huffman solution. Both systems begin by creating one node for each character, but they build the tree differently. The Huffman method starts from the bottom and builds up, while the Shannon-Fano method builds down. Here's a short summary:

1. For each character x, create one node n_x.
2. For each node, attach a null string, $c(n_x)$, that will hold the final code for each character.
3. Put all of the nodes in one set N.
4. Recursively apply these steps to the set N.
   (a) Split N into two sets, N_1 and N_2, so that the sum of the probabilities of the characters in N_1 and N_2 are as close as possible.
   (b) Add a '0' to the end of the $c(n_x)$ for all n_x in N_1. Add a '1' to the end of the $c(n_x)$ for all n_x in N_2.
   (c) Recursively apply this function to N_1 and N_2 separately.

This solution will create a string of codes for each character that will be similar to the Huffman codes. In fact, the codes may be exactly the same. It all depends upon how the set N is split into N_1 and N_2. This process is fairly difficult to do optimally, and the fastest solution is to use an approximation to find the best split. One of the easiest ways to approximate a good split is to use the same algorithm that is used to build the Huffman trees. This is why the Huffman algorithm is the better-known solution and the one that is used most frequently.

The section on page 20 describes how to encode and decode Huffman codes. These also work here.

Some call the trees built in the Huffman algorithm extended binary trees *instead of* Huffman trees.

Once the tree is complete, the coding and decoding processes are the same for Shannon-Fano and Huffman coding. The variable length codes developed by this algorithm are also free of prefix ambiguity, so the same algorithms given before will suffice.

2.3 Entropy and Information Theory

Both Huffman and Shannon-Fano codes are closely related to the theory of information developed by Claude Shannon in the 1940s [Sha48, Sha59]. This theory provides a statistical way to measure information that is, in at least one sense, more insightful than traditional ways. In the most abstract sense, the theory provides a mathematical way of measuring the news value of a particular bit of information. If a data pattern is particularly likely, then it has less information in this framework than a pattern that is rare and unlikely. The theory works like a news editor making judgments over what stories deserve front-page headlines. A weather report

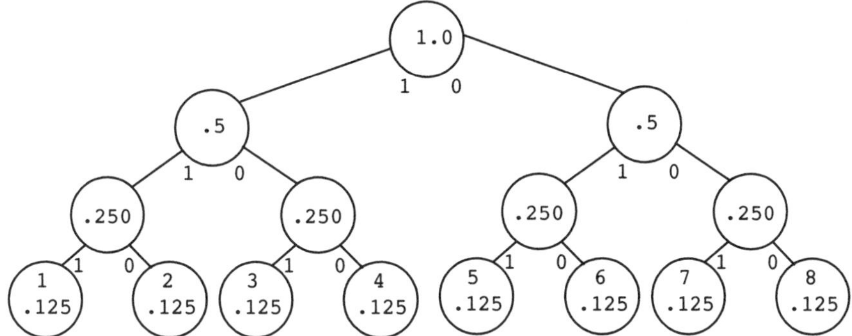

Figure 2.3: A Huffman tree constructed for an alphabet with eight characters that appear with equal probability. All end up with codes of the same length.

saying that it will be 72 degrees Fahrenheit and sunny in Los Angeles is not news, but a report predicting three feet of snow there would be news.

Another measure of information is Kolmogorov Complexity, *which is discussed on page 12.*

Shannon used the term *entropy* to encapsulate the measure of information. The term is used widely in physics to describe the amount of order or disorder in a system. In this case, Shannon argued that a system with a high degree of disorder was also one that contained a great deal of information.

Shannon offered this equation as a measure of information:

$$\sum_{x \in A} \rho(x) log_2 \frac{1}{\rho(x)}.$$

The logarithm in base two of the reciprocal of $\rho(x)$ approximated the best count of the number of bits it would take to represent that character and the sum computed the weighted average.

Some other books on information theory are [Abr63, CT91, Gal68, Ham80]

Consider the example of an alphabet $A = \{1, 2, 3, 4, 5, 6, 7, 8\}$, where $\rho(x) = \frac{1}{8}$ for all $x \in A$. The entropy of this file is $8 \times \frac{3}{8} = 3$. Figure 2.3 shows a Huffman tree built to illustrate this compression technique. Notice that each character in A will end up with a three-bit code because the tree is three levels deep. If a file with these statistics were compressed with Huffman compression, then it would take up three bits per character, an amount that is equal to the amount predicted by the entropy equations.

Now consider the four-character alphabet that leads to the Huffman tree con-

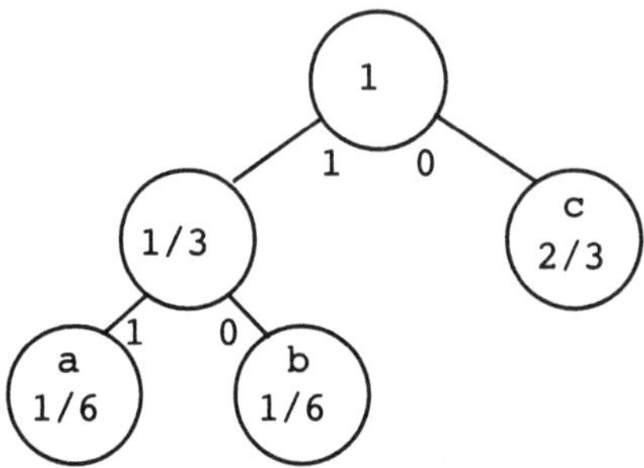

Figure 2.4: A Huffman tree for a simple file that does not compress the file to the same number of bits as the entropy figure predicts.

structed in Figure 2.2. The entropy of this file is:

$$\frac{log_2 2}{2} + \frac{log_2 4}{4} + \frac{log_2 8}{8} + \frac{\log_2 8}{8} = 1.75.$$

This is also the average number of bits used per character to compress a file with the same distribution of characters.

The entropy of a data stream and the number of bits used to compress the data stream with Huffman encoding are not always the same. In fact, they are only equal when each of the logarithms turns out to be a perfect integer, that is, when all of the probabilities are negative powers of two. This is the only time when the number of bits used per character is exactly equal to the weighted entropy contributed by that character. The rest of the time, the fractions prevent the Huffman coding from doing the best possible job.

Arithmetic coding described in Chapter 4 exploits this limitation to get better compression.

Consider the example illustrated with the Huffman tree in Figure 2.4. In this case, one character occurs two-thirds of the time, and the other two appear one-sixth of the time. The entropy of this stream is $\frac{2}{3}log_2\frac{3}{2} + \frac{2}{6}log_2 6$, which is about 1.2516. If a file is compressed with the Huffman tree in Figure 2.4, however, two-thirds of the characters will be converted into single-bit codes, while the remaining third will get two-bit codes. This means $\frac{4}{3}$ bits will be used per character, which is more than the entropy.

Shannon postulated that the entropy of a data stream is the minimum number of bits per character that can ever be achieved in compression. That is, it is the maximum amount of compression possible for a file. This can be seen in an informal way by noticing that files compressed to their maximum entropy will be essentially random.

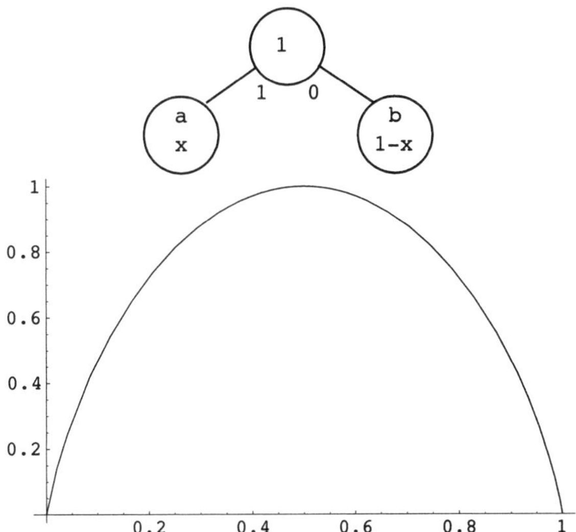

Figure 2.5: A simple Huffman tree produced for a two-character alphabet where one character occurs with probability x and the other occurs with probability $1 - x$ ($0 \leq x \leq 1$). The graph at the bottom of the figure shows the entropy for this file as a function of x.

Consider the result of compressing a file using the tree in Figure 2.3. Each of the eight characters is equally likely to occur, which means that each of the eight three-bit codes from '000' to '111' is also equally likely to emerge in the compressed file. The number of zeros and ones in the final compressed file should be equal.

The same effect happens when the file that produced the tree in Figure 2.2 is compressed. This is more difficult to see in this example, but it can be proven with a bit more work. Instead of proving this in detail, consider this simpler example that illustrates how Huffman codes fail to generate optimal compression when the probabilities are not negative powers of two.

Figure 2.5 shows a simple Huffman tree constructed for a two character alphabet. In this case, one character ('a') occurs with probability $\rho(a) = x$, and the other occurs with probability $\rho(b) = 1 - x$. The Huffman tree remains unchanged despite the value of x, which means that the weighted average number of bits used per character will be $x + (1 - x) = 1$. The entropy of this file, however, is plotted in the graph on the right of Figure 2.5. The maximum of 1 only occurs when $x = \frac{1}{2}$. Notice how the entropy drops to zero when either character dominates the file. In this theory, there is no information in a file, no matter how long it is, if only one character is guaranteed to occur. [1]

This example is a good constructive proof of two points. First, perfectly random two-character alphabets have maximal entropy. Second, Huffman codes fall short of achieving maximal compression when the probabilities are not negative powers of two.

Both of these principles extend to general files with arbitrary alphabets, and these principles can be proven with more rigorous detail by using recursion.

It is interesting to compare the Shannon model to the basic model that counts the number of bytes in a file. As this section has already shown, the number of bytes in a file might not be the best measurement of the amount of information in the file. If it were a true measure, then the file should not be compressible. If a file like the one that generated the Huffman tree in Figure 2.3 were compressed, it would go from $8n$ bits to $3n$, bits assuming that each uncompressed, initial character was stored as eight bits using the standard ASCII code. But the amount of the entropy has remained unchanged. It went from three bits per n 8-bit characters to one bit per $3n$ 1-bit characters. In this case, the word "character" is a slippery one.

If Shannon information theory gives a more accurate measurement of the amount of information in a file, why isn't it more widely used? The main problem is that it has many different limitations, and it is often difficult to compute perfectly.

Consider a file that just consists of the eight letters 'abcdefgh' repeated n times. In this file, each of the eight characters occurs with equal probability. This means that the entropy of the file is $3n$, and the Huffman tree from Figure 2.3 can be used to reduce the file from $8n$ bits to $3n$ bits by replacing each character's ASCII representation with its three-bit Huffman code.

But better compression can be achieved. Imagine that we redefined the alphabet, A, to consist of two 'characters': abcd and efgh. Both of these 'characters' occur n times with equal probability. The Huffman tree from Figure 2.5 can be used to compress the file to be n bits long. This is much better than the entropy estimate predicted was possible, but it is clearly possible only after breaking one of the unspoken assumptions about the alphabet. The entire discussion of the model and Huffman codes up to this point assumed that the characters were all atomic and independent. There was no way to aggregate them or pick up patterns that

[1] Clearly the length of a file containing only one character is still a piece of information. Shannon's theory was originally conceived to model data streams on radio channels. These are, essentially, files without end. In this context, measuring such a stream of unchanging characters as zero is clearly fair.

might influence things.

These problems are also opportunities for better compression. Before this point in the chapter, the probabilities were assumed to be independent of each other. That is, each character occurred with the probability, no matter which characters came before or after it. This is clearly not the case in the English language, where a 'q' is almost always followed by a 'u'.

In this example, the entropy can be calculated with this equation:

$$\sum_{a \in A} \sum_{b \in A} \rho(a, b) log_2(\frac{1}{\rho(a, b)}),$$

where $\rho(a, b)$ means the second-order probability that b will follow a. A collection of Huffman trees can be created with one tree for each character in the alphabet. After a letter is encountered in the file, the Huffman tree corresponding to that letter is used to compress the next character. If the probabilities are negative powers of two, then the Huffman compression will generate the same compression suggested by the entropy calculation. Otherwise, it will fall short.

The basic statistics are known as first-order *statistics. nth order statistics compute the probability that a character will occur based upon the $n - 1$ previous characters.*

This scheme does a better job of compressing the English language. Some letter combinations are quite rare and others are more common. For instance, 'g' and 'h' are both relatively common in the English language, but a 'g' is less likely to occur after a 't' than is an 'h'. The Huffman tree constructed explicitly for letters following 't' will take this into account and do a better job compressing the file.

Still, there are problems with this scheme. The English language has deeper interrelationships between letters. For instance, an 'e' is common after a 'th', but rare after an 'gh'. Why not create an even better compression scheme by creating one Huffman tree for each *pair* of letters. The compression will still be better, but it comes at a cost. A copy of each tree must be available at decompression time. This might not be a problem with fixed systems that are continually processing large files with the same statistical patterns, but it can be wasteful if the trees must accompany the compressed file. The extra overhead can hurt compression.

Peter Fenwick shows a way to encode files by ranking the likelihood that a particular letter will follow. The code is the position in this ranking. This is quite similar to building a separate Huffman tree for each character.

There are still deeper problems with this model. Even nth order statistics fail to capture all of the nuances of the English language, no matter how large the value of n. For instance, in well-written English text, the noun must agree with the verb. That is, if a plural noun is used, then a plural version of the verb must also be used. Raw statistics cannot capture this basic pattern, because the noun and the verb are often separated by several words.

Shannon worked around these problems by simply proposing that the n-th order entropy of a stream of data was computed by giving a person access to the previous $n - 1$ characters and letting him or her predict the next character. A person could pick up more complex rules, like the one about nouns and verbs, as n got large enough.

This solution may help smooth over the theory of information, but it doesn't help a computer program trying to create a good Huffman tree for compression. At this point, the relationship between Shannon's theory and these models starts to diverge. There are increasingly complicated ways to estimate the probabilities of a stream of data, but the Huffman coding algorithm described in this section is not equipped to take advantage of them.

2.4 Character Grouping Schemes

The first sections of this chapter assumed that all of the characters in a file were independent of each other. The previous section on entropy and Shannon's theory of information introduced the notion that the concept of a character could be flexible. Sometimes, more compression can be achieved by aggregating several characters together in the Huffman tree.

The rest of this chapter will use the term *token* instead of *character* because it better encapsulates the notion that the file might consist of multicharacter units that are treated as one.

The easiest way to understand how this might occur is with the English language and the letters 'q' and 'u'. A basic scheme for computing the Huffman codes would put a 'q' in one node and a 'u' in the other. If the file contained English text where every 'q' was followed by a 'u', then it makes little sense to have a separate node for 'q' and 'u'. The best solution is to combine them into one token that gets its own node in the Huffman tree.

This solution always makes sense if one token is always followed by another. There is no need to spend any extra bits to encode the trailing token.

The decision gets more complicated when multiple tokens are involved. Consider the case where one token, a, is always followed by one of two different tokens b or c. It makes sense to remove the token a from the alphabet and replace it with two new tokens ab and ac if this is true:

$$
\begin{aligned}
\rho(a)log_2\frac{1}{\rho(a)} + \rho(b)log_2\frac{1}{\rho(b)} + \\
\rho(c)log_2\frac{1}{\rho(c)} &\leq \rho(ab)log_2\frac{1}{\rho(ab)} + \\
&\quad \rho(ac)log_2\frac{1}{\rho(ac)} +
\end{aligned}
$$

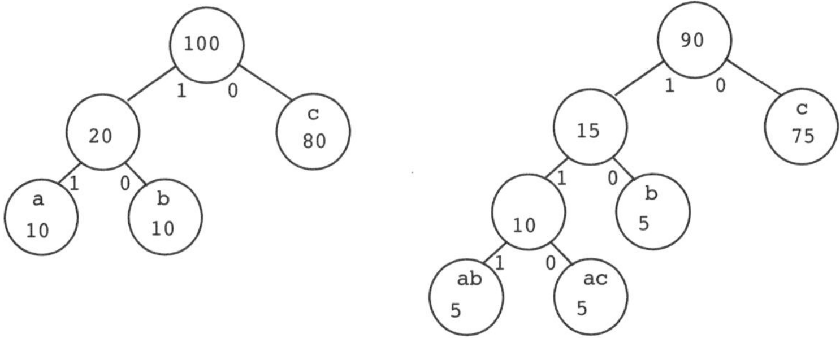

Figure 2.6: Two Huffman trees computed for a simple file. The circles hold the tokens and the number of times that the tokens occur. This illustrates how it can make more sense to aggregate tokens.

$$(\rho(b) - \rho(a))log_2\frac{1}{\rho(b) - \rho(a,b)} +$$
$$(\rho(c) - \rho(a))log_2\frac{1}{\rho(c) - \rho(a,c)}.$$

This is rarely the case in theory. It can still be true, however, that it makes sense to aggregate tokens. For instance, Figure 2.6 illustrates an example of a file where there are three initial tokens: a, b, and c. These occur 10, 10, and 80 times in the file. In this particular case, the token 'a' is followed by a 'b' half of the time and a 'c' the other half. This pattern means that it is possible to consider replacing the token 'a' with two tokens 'ab' and 'ac'.

If the Huffman tree on the right is used to compress the file, then $1 \times 80 + 2 \times 10 + 2 \times 10 = 120$ bits will be used. If the tree on the right is used, then $1 \times 75 + 2 \times 5 + 3 \times 5 + 3 \times 5 = 115$ bits will be used. Five bits can be saved by doing this aggregation.

This solution does not always work. Figure 2.7 shows how the technique fails to generate any extra savings. The tree on the left is built for four tokens that occur with equal probability. Each of the characters also follow each of the others with equal probability. The tree on the right shows the effect of removing the node for the token '1' and replacing it with the four possible two-character tokens beginning with '1'. There are four possible solutions ('11','12','13','14'), because each character occurs with equal probability in the mix.

This solution is equivalent to merging two trees. The top tree is the standard tree built from statistics about the entire file. The bottom tree is the one built from statistics about what comes after the character '1'. In this case, this bottom tree is

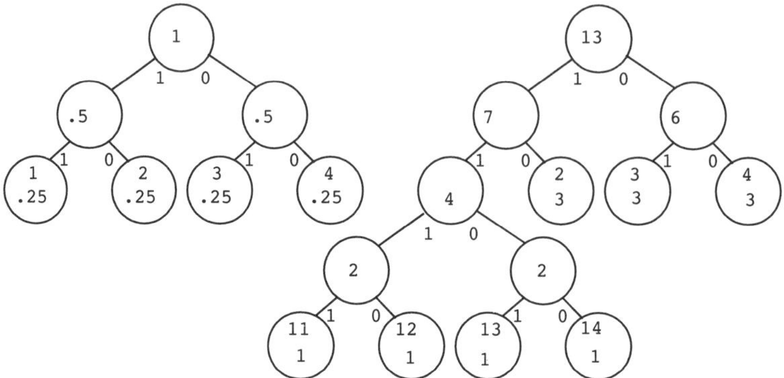

Figure 2.7: This figure shows two Huffman trees before and after the tree was merged with itself. This particular action did not save any of the disk file because the same number of bits were needed before and after.

as full as the top. There is no difference between the statistics over the whole file and after the character '1'.

No extra bits will be saved in this process because it will take two bits to encode tokens before and an average of two bits per character afterwards. The tokens in the right tree take two bits if they contain a single character and take four bits if they contain two characters.

On occasion, merging trees like this can force them to be redrawn. Figure 2.8 shows a similar situation to Figure 2.6. There are three tokens, 'a', 'b', and 'c'. It is observed that an 'a' is followed by a 'b' four times in the file and a 'c' six times. So the decision is made to remove the 'a' token and replace it with two new tokens 'ab' and 'ac'. This is equivalent to merging the tree with one constructed with the statistics for tokens following 'a'.

This new tree, shown on the right, is not balanced correctly. Notice that the branch containing the 'c' token will occur nine times in the file, but the branch containing the 'ab' and 'ac' tokens will occur ten times. The proper solution is shown in the bottom tree. This can be accomplished by simply swapping the two subtrees.

The top left tree requires 55 bits to represent the file: 15 for each 'c', 20 for each 'a', and 20 for each 'b'. The right tree requires 51 bits ($c = 9, b = 12, ab = 12, ac = 18$). The bottom tree takes 50 bits overall, using two bits for each token.

These examples are drawn with smaller than usual samples. In most of these cases, it doesn't pay to compress the file because the amount saved is eaten up by

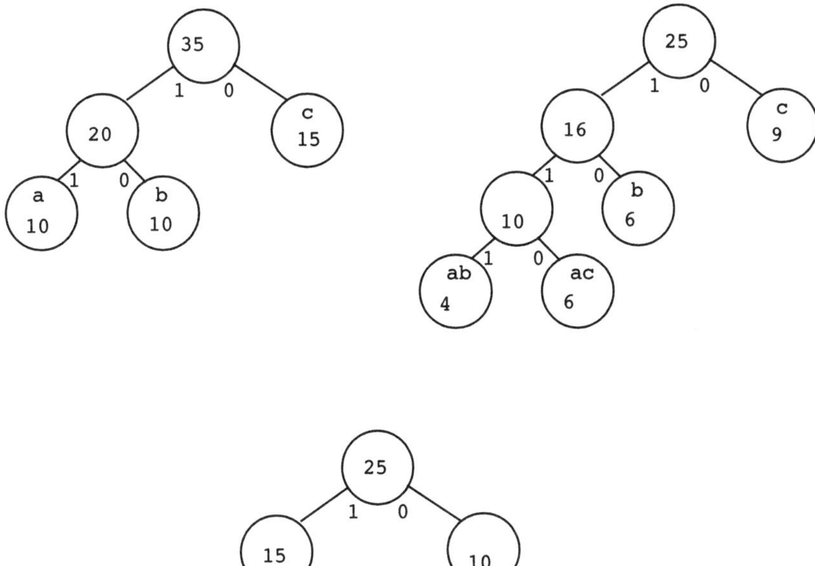

Figure 2.8: Here are three trees. The first is a simple Huffman tree built for three tokens. The second shows it after it is merged. The bottom tree shows the correct balance for this tree to produce the right compression.

the overhead.

A Basic Algorithm for Merging

The last section has presented many examples for how to merge tokens and grow trees. In some cases, several bits were saved. In other cases, none were saved. Here's a basic algorithm that shows how the process can be repeated.

1. Calculate second-order statistics. Identify a token that is always trailed by only a few tokens.

2. Create new nodes that merge each of these tokens.

3. Remove the old token, and replace it with a tree created with just these tokens. The count of all of the merged tokens in the new tree will be the same as the count of the old token being replaced.

Chapter 5 shows similar procedures for mutating the Huffman tree during the compression process.

4. Subtract the occurences of the following tokens from their rightful spot.

5. The tree may be incorrectly balanced. Figure 2.8 shows an abstract situation before and after three branches are reordered. The new tree can be rebalanced by starting at the bottom of the new merged nodes added to the tree and working up the tree looking for places where the tree is unbalanced. If this is the case, then a swap should be made. This process should be repeated at each node along the way until the root is reached.

Programmers using this merging method must keep track of the amount of space used in the overhead. Adding new nodes will grow the overhead, and the merging should only be attempted if it will shrink the overall size of the file. This amount will vary depending upon the method that you use to store the Huffman tree.

2.5 Conclusion

This chapter described how to use basic statistical information about the contents of a file to come up with a variable length code for each item in the file. Recoding the file with these new codes will shrink the overall file, because short codes are assigned to the most common tokens in the file and longer codes are left for the least common.

Some patents on ideas from this chapter are found beginning on page 180.

The chapter also provided some theoretical insight into the algorithm by describing Shannon's definition of information entropy. The two coding algorithms provided in this chapter are based upon this work, and it provides a good estimate for the best compression that can be achieved.

Still, this chapter also explained how the theoretical model can fail to provide adequate understanding. There is no clear way to define how to break up the file into tokens used to build the tree. While character-based tokens are natural and easy to use, they might not be the most efficient. The chapter provided some ways to merge tokens and further shrink the file.

There are still many features that are not described in depth. One common solution is to create one token for each word in the file. This a cross between the Huffman techniques in this chapter and the dictionary-based solutions described in Chapter 3. This solution works well if many of the words are used more than once, but it begins to fail if there are many words used only once in the file. This is because the word must be stored once in the overhead table and once in coded form in the compressed file. This is bound to be lengthier.

Many other algorithms use the basic techniques described in this chapter to help with compression. The image compression algorithms like JPEG (Chapter 10) and MPEG (Chapter 11) routinely use Huffman compression to add an extra level of compression to the data used to represent the image.

Chapter 3

Dictionary Techniques

One of the most common solutions for compressing data is to create a list of the most common words or phrases in a document and then replace each of these blocks of letters with a short number representing the position in the dictionary. This can be quite powerful if the dictionary is the right size and there are plenty of common patterns in the file.

The best-known version of these dictionary schemes, which are sometimes called *substitutional compression*, is called "Lempel-Ziv" after Jakob Ziv and Abraham Lempel. The algorithm, first described in 1977 [ZL77], is used in many different compression programs in many different versions. Some of the latest versions have been patented, and the patents have been the basis for some bitter battles over rights and ownership.

Appendix A describes some of the major patents in compression and provides a summary of the effects.

Many of the versions differ in only minor ways, but these differences are occasionally crucial for producing better compression.

The basic points of contention are how to create the dictionary, how to ship the dictionary to the decompression routine, and how to store the items from the dictionary. Several people have offered their own versions over the years, and this chapter will try to present some of them in enough detail to encode them.

The basic engineering challenges in designing these dictionary-based solutions are, in no particular order, as follows:

Dictionary Transmittal All of the basic archiving schemes do not transmit the dictionary in a separate package. They send the file in normal form and insert special escape characters that indicate that a special phrase should be inserted into the dictionary. Both the compressing program and its matching decompression program must synchronize how they handle storing these phrases in the dictionary.

In some fixed compression situations, it may be possible to distribute copies of the dictionary in advance. This can save substantial complexity in the algorithm and also allow for more sophisticated analysis of the entries in the dictionary. Most of the algorithms in this chapter are designed to adapt themselves to the data as it is being processed and need to be substantially rewritten to take advantage of these possibilities.

Dictionary Storage There is a great variety in the manner of storing the dictionary. Some simply maintain a window of the last n characters in the file and use this as a default dictionary. This takes little work and also implicitly refreshes the dictionary. Other schemes maintain complicated data structures that constantly rebalance themselves and purge themselves of phrases that aren't frequently used. Some use moderately complicated schemes.

Code Word Length Some dictionaries use a fixed size, such as 4096, and limit entries to this count. This means that all of the code words have a fixed size, such as 12 bits, but it also requires purging the dictionary of phrases. Others allow the code words to grow in size, but this trades the complexity of purging the dictionary for the complexity of identifying the codewords.

New Phrase Identification Some algorithms are cautious and allow the list of phrases to grow only one by one. Others try to group phrases together aggressively. The cautious approach works well if many longer phrases only occur once. They are not added to the dictionary only to be removed later. The aggressive solution works better if there are many long phrases that occur frequently.

3.1 Basic Lempel-Ziv-Welch

Many programs use a version of Lempel-Ziv created by Terry Welch in 1984. This version is pretty simple and easy to code , so it is frequently included in many simple compression schemes. It is also relatively good, although it takes its time building up the dictionary.

The compression algorithm can be summarized with this pseudocode:

```
word="";
while not end of file;
   x=read next character;
   if word+x is in the dictionary
       word=word+x;
   else
       send the dictionary number for word
       add word +x to the dictionary
       word=x
 end of while loop
```

In the pseudocode, the addition sign ('+') stands for concatenating a string variable (word) with a character (x). In this algorithm the dictionary begins with 256 entries numbered 0 through 255. These are the single bytes, which usually correspond to ASCII characters. New entries start at number 256 and go up. Some versions will prune the number. Others expect the decompression algorithm to recognize that the table is growing larger and allocate more space to each code.

The decompression loop is also just as simple to code:

```
 read a character x from compressed file;
write x to uncompressed version;
 word=x;
while not end of compressed file do begin
    read x
    look up dictionary element corresponding to x;
    output element
    add w + first char of element to the dictionary
    w = dictionary element
 endwhile
```

The two algorithms maintain the same version of the dictionary without ever transmitting it in a separate file. There is still some overhead, however, because the compressed version contains only dictionary entries, not individual data elements, and these elements, are all larger.

An Example of LZW

The best way to understand how the Welch version of Lempel-Ziv works is to track its behavior on a string of data. Here's how it works on "BABY AT BAT".

Input Character (x)	Output Token	New Entry	word value
B	none	(none)	B
A	B	257 = BA	A
B	A	258 = AB	B
Y	B	259 = BY	Y
space	Y	260 = 'Y '	space
A	space	261 = ' A'	A
T	A	262 = AT	T
space	T	263 = 'T '	space
B	space	264 = ' B'	B
A	none	none	BA
T	257	265 = BAT	T
none	T	none	none

A data structure for holding the dictionary entries in this example is shown on page 44.

This example shows how the algorithm builds up a fairly substantial table at the beginning before it happens upon a pattern that it could compress. This example only shows one place where any compression occurs—at the end of the table where the letters 'BA' are replaced by the code 257. The algorithm does not have any way to make use of the fact that the letters 'AT' are also in the table.

Here's another example with the phrase "BOB BOOTS BOOMBOX BOATS".

Input Character (x)	Output Token	New Entry	word value
B	none	(none)	B
O	B	257 = BO	O
B	O	258 = OB	B
space	B	259 = 'B '	space
B	space	260 = ' B'	B
O	none	none	BO
O	257	261 = BOO	O
T	O	262 = OT	T
S	T	263 = TS	S
space	S	264 = 'S '	space
B	none	none	' B'
O	260	265 = ' BO'	O
O	O	266 = OO	O
M	O	267 = OM	M
B	M	268 = MB	B
O	none	none	BO
X	257	269 = BOX	X
space	X	270 = 'X '	space
B	none	none	' B'
O	none	none	' BO'
A	265	271 = ' BOA'	A
T	A	272 = AT	T
S	T	273 = TS	S
none	S	none	none

This run of the algorithm does a better job of finding several places to compress data. Still, the process of building up the dictionary can seem slow here. The algorithm is not able to use the information it gained about string 'BOO' which was stored in entry 261 later. When it encounters the string a second time, it is already building up a version that begins with a space. This ends up as 265 and 266.

While this process may seem quite slow in these short example, it can still be effective. Most of the common letter pairs are in the dictionary after the first several hundred bytes. Most of the common short words like 'the' are also part of the table by the fourth or fifth time they occur. Once this point is reached, substantial savings can begin accruing. If an n letter word occurs m times, then it will be completely compressed after the $n + 1$th time it occurs. The extra one unit is for the trailing space. This can be a substantial savings.

The big advantage of this algorithm is that the dictionary is built up on the fly. There is no need to transmit it as overhead. Nor is there any reason to store words in the dictionary that don't occur in the document.

There is still overhead to the method. The values being written to the output file are not bytes representing ASCII values, they're tokens representing entries in the dictionary, and these tokens that are being shipped are not one byte long. Numbers like 260 need at least nine bits to hold them. This description has not specified how these tokens are written to the output file, but there are two basic mechanisms.

The simplest approach chooses a fixed power of two to be the size of the dictionary and uses n bits to write each token. One common solution uses $n = 12$ and places $2^{12} = 4096$ entries in the dictionary. This means that 12 bits are used for each token. Clearly, the first example from page 37 would not save anything. Eleven bytes taking up 88 bits are converted into 10 tokens taking up 120 bits. The example on page 38 does somewhat better. Twenty-three letters (184 bits) are compressed into 18 tokens (216 bits).

A more sophisticated approach uses only enough bits to represent the dictionary's current size. If there are between 2^n and $2^{n+1} - 1$ entries in the dictionary, then n bits are used. In both examples, the size of the dictionary does not grow above 511 entries, so nine bits suffice. In the first case, 88 bits of ASCII are converted into 90 bits of dictionary entries. In the second case, 184 bits of ASCII are compressed into 162 bits of tokens.

Both of these examples began with all 255 ASCII values in the dictionary already. Some versions start with only the first 127 values, because these are the only ones that normally appear in text. This approach is good for text, but worse for other data streams with a full range of byte values.

Here's a table showing how the decompression algorithm works on the first example beginning on page 37:

Input token (x)	Output Letter	New Entry	word value
B	B	(none)	B
A	A	257 = BA	A
B	B	258 = AB	B
Y	Y	259 = BY	Y
space	space	260 = 'Y '	space
A	A	261 = ' A'	A
T	T	262 = AT	T
space	space	263 = 'T '	space
257	BA	264 = ' B'	B
T	T	265 = BAT	T

The dictionary can be built up in synchrony with the compression algorithm, and this allows the process to work successfully.

3.2 Simple Windows with LZSS

James Storer and Thomas Symanski are responsible for another version of Lempel-Ziv's algorithm that is often encountered. This version does not attempt to keep a dictionary filled with the various pairs, triples, and other combinations that are frequently seen. It just maintains a "moving window" of the file and uses this as a dictionary. Instead of referring to "entries" in the dictionary, it refers to the position in this moving window.

The algorithm is pretty simple. The last n bytes of the file are maintained as a window. n is usually some power of two to make it easy to represent as $log_2 n$ bits. The algorithm keeps another window of the next m bytes to be compressed. In each round, it looks through the entire n byte window for the best match of the m bytes. If it can find it, it will transmit the offset and the length instead of the letters.

Here's the algorithm in pseudocode:

```
past = nil \ This is the n byte window.
future= first m bytes of the file
do
   search through past for occurrence of future.
   if the length of the match is greater than a threshold
       transmit a pair with (offset,length)
       remove length bytes from future
       add them to the end of the past
       get length new bytes for the end of future
```

```
   else
      transmit the first  byte from the future.
       remove it from the future
      add it to the end of the past
      load new byte
    end
 while bytes available
```

Clearly the most time-consuming part of this algorithm is searching through the window of the last n bytes. The larger this value is, the longer the algorithm will take to compress the file. The size of n does not affect the decompression time. Some data structures can help this process, but they are beyond the discussion at this time. They are often so involved that it is better to classify the systems as completely new algorithms.

One of the biggest problems a programmer must solve is how to represent the tokens produced by this compression algorithm. One simple solution is to designate one value to be an escape code. A good choice may be 255 because it doesn't appear too often. An escape code signals that the next values will represent an offset and a length, not raw data. What happens if the escape code is part of the file and needs to be transmitted? In this case, the code may be followed by a special offset and length pair that mean the code itself.

There are several possible approaches to storing the offset and the length. A simple system allocates one byte for the offset and one byte for the length. This may be easier to work with, but it can be wasteful. In most cases of overlap will be relatively small. Another solution is to block off k bits and allocate 4 or 5 to represent the length and $k-4$ or $k-5$ to represent the offset. This depends upon the size of the window.

An Example with LZSS

Here's a table showing the various steps involved in compressing the phrase "BOB BOOTS BOOMBOX BOATS" from the example of LZW shown beginning on page 38. This example will set the size of `past` and `future` to be eight bytes long and set a minimum threshold of two characters for there to be compression. The column labeled "transmitted" shows the values transmitted in the compressed file. If there is a single character, then the character alone is sent. If there is a pair in parentheses, then it is sent as an offset and length.

past	future	transmitted
(nil)	BOB BOOT	B
B	OB BOOTS	O
BO	B BOOTS	B
BOB	' BOOTS B'	space
BOB	BOOTS BO	(0,2)
BOB BO	OTS BOOM	O
BOB BOO	TS BOOMB	T
BOB BOOT	S BOOMBO	S
OB BOOTS	' BOOMBOX'	(2,4)
OOTS BOO	MBOX BOA	M
OTS BOOM	BOX BOATS	(4,2)
S BOOMBO	X BOATS	X
' BOOMBOX'	' BOATS'	(0,3)
OMBOX BO	ATS	A
MBOX BOA	TS	T
BOX BOAT	S	S

How much space is saved in this example? It depends upon how the offset and length values are packed. Three bits are fine for both of them. If the value 255 is used as the escape code, then the next byte can hold the offset and length together. (255 followed by 255 signifies that the escape code is found in the file itself.) This requires two bytes to store each offset and length pair. Two of the codes ([4,2] and [0,2]) save no extra space, but don't cost any either. Two codes ([2,4] and [0,3]) save three bytes together.

Another option in this case is to pack the data even smaller. Most ASCII files never have values above 127. If this is the case, then the eighth bit can be used as a flag to signify the presence of an offset-length pair. If the values are less than 127, then they should be interpreted as raw data. If the values are greater, then the last six bits should be thought of as an offset-length pair. In this case, seven bytes would be saved.

This approach, however, has some overhead. There is an extra bit attached to each value that signifies whether it is an offset-length pair or raw data. Twenty-three bits could merely be saved by storing seven instead of eight bits of each ASCII value.

Another option, described by Adisak Pochanayon, is to include metatags every so often describing whether the next several bytes will be raw data or length-offset pairs. In his example, he uses the eight bits in a byte to specify whether the next eight values are raw data or offset-length tokens. This is repeated throughout the algorithm. This approach is valuable because it separates the flag bit from the data.

This may be useful if the length and offset pairs are odd sizes. In the example above, they might only be six bits long, and this could be reduced to five because there is never a length longer than four meaning that the length could be packed into two bits.

3.3 Coding Notes

The biggest problem in implementing a variation on this Lempel-Ziv algorithm is finding a quick way to search the dictionary. Most of the basic data structures can be of some help. Here's a list of some solutions and their advantages and disadvantages:

Hash Tables A large hash table consists of a function $f(x)$ that converts some word or string of letters (x) into a position in a table. This function is normally fairly random and might look something like $x[0]^3 + x[1]^2 + x[2] + x[3]^2 mod\, n$, where $x[i]$ is a numerical representation of the ith character in the string. n is the length of the table.

Hash tables like this make an easy way to look up values. There is one computation, the hash function, that points to a table entry that will probably contain the dictionary entry and its value in the dictionary.

The disadvantage is that hash functions often lead to *collisions* when two words produce the same value in the hash table. In this case, it might be preferable to build a linked list of dictionary entries emerging from the table. This makes it easy to put multiple dictionary entries at the same location. This is often useful in schemes like LZW that can have arbitrarily long dictionary entries. Hash functions may only depend upon the first k letters in x.

A common solution is to make the hash table relatively small and count upon the linked lists to grow. The hash function might key off only the first three or four letters.

In the case of plain LZ77 or LZSS solutions, the linked list growing off of each hash table entry should point to all offsets in the window with that letter combination.

Matrix and Linked List One solution, which is just a variation of the hash table proposed above, builds a simple two- or three-dimensional array with pointers to linked lists. In the case of LZW, the linked lists contain all dictionary entries beginning with the first two or three characters. In the case of LZSS, the lists contain pointers to all of the offsets in the `past` window. The two- or three-dimensional array will be keyed by the first two or three letters in the dictionary entry. In essence, the array lookup process is just a simple hash function.

This approach may waste space because many of the entries in the two- or three-dimensional array will be empty, while others will be overloaded. Modern computers, however, have ample memory, so this scheme can be fairly useful. Still, three-dimensional arrays can be prohibitive if the value of each dimension ranges over the full 0 to 255 expressed by eight bits. One solution is to keep only the bottom six bits as an index for the arrays. These contain most of the information in ASCII data, so there will be few collisions in text files. This means that the array must hold 2^{18} instead of 2^{24} entries — a substantial savings by a factor of 64. Still, collisions will be a problem and must be anticipated by the code used to search the linked lists for the proper dictionary entry.

Tries or Alphabetic Trees Another popular solution is to store the dictionary entries in alphabetically indexed trees, often called *tries*. These are trees where the descendants are kept in alphabetical order.

Figure 3.1 shows a set of tries corresponding to the run of LZW described beginning on page 37. Each node contains a pointer to a list of descendants kept in alphabetical order. In this illustration, there aren't many branches in the tree structure because there aren't many entries in the dictionary, but these trees can grow quite bushy.

Each node contains the number of the dictionary entry corresponding to the string of characters from the root of the tree to that point. In Figure 3.1, the value 265 represents the dictionary entry for BAT.

This data structure works quite well for the LZW approach to maintaining a dictionary. If no entry is found for a particular word, then the next character can be added at this point. When the dictionary is grown one letter at a time, this approach is easy to implement. If dictionary entries are merged, then the approach is less useful.

This solution is also less useful for LZSS because the process of deleting entries requires removing several nodes. This is not too hard, but it is more onerous than simply removing one entry from a table.

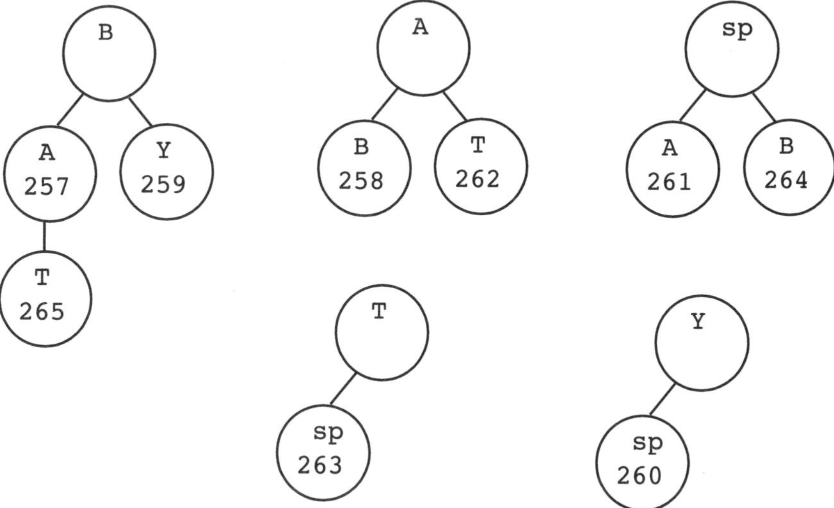

Figure 3.1: The tries built to store the dictionary entries from the run of LZW described on page 37. Each tree is alphabetically sorted. If there is a number in the node, then it corresponds to a dictionary entry.

It makes sense to mention that the raw buffer from the LZSS algorithm used to store the n bytes of the `past` string and the m bytes of the `future` are often combined into one long round buffer. A pointer into the circular array indicates the distinction between `past` and `future`. The approach has two distinct advantages. The first is that letters do not need to be stripped off and copied during every update of the two variables. The movement of bytes from `future` to `past` can be accomplished by moving a pointer into the array.

The second advantage is that no updates need to be made to any secondary data structure built to speed up searching. If a hash table or array has an entry saying that the letters 'Hi' are found at position 259 in the array, then they will stay there until they are removed from the buffer.

3.4 Variations

There have been a wide variety of modifications made to Lempel-Ziv that go beyond the two major versions described above. The most significant change may be how the dictionary code words are stored. One solution is to use some frequency analysis to find the most commonly substituted dictionary terms. This information is then used with a Huffman compression algorithm to compute variable-length code words for the dictionary entries.

Chapter 2 shows how to create variable-length code words based on the statistical profile.

This solution can be attempted in two ways. The first is to run a second pass over the tokens generated by the dictionary compression algorithm and use the Huffman compression to reduce the final size even more. This works well with LZW with an ever-expanding dictionary.

LZSS does not work well with this scheme because the offsets for popular words are often changing. This problem can be reduced by using a circular buffer with fixed positions for the offsets and the lengths.

Adaptive solutions are described in Chapter 5.

Another solution is to use adaptive Huffman coding.

3.5 Commercially Available Standards

There are a wide variety of programs designed to compress files using Lempel-Ziv-like compression algorithms. Many of them include their own special data structures or tweaks. Some have publically distributed source code that you can use to read or write the files.

The LZW system with a dictionary built on the fly is used in the *arc* files which are sometimes known as *pkarc* files. The popular UNIX *compress* program, which generates Z files, also uses this algorithm.

The LZ77 system, which represents frequent words as distance and offset pairs, is also found frequently. The *arj* system, *sqz*, *uc2*, *lzh*, *zoo*, and *lha* all use the system and add a pass of Huffman coding for good measure. The most common system using this approach, however, is probably the *zip* file format which is commonly found throughout the Internet thanks to the popularity of the `pkzip` software package. Versions 2.04 and later use this algorithm along with the popular free GNU version known as *gzip*.

GNU Zip or GZIP

The GNU zip algorithm is a mixture of dictionary and statistical approaches developed by Jean-Loup Gailly and Mark Adler. The source code, descriptions of the algorithms, and implementations for various machines can be found at the website `www.gzip.org`. The software is well-regarded and freely distributable. The authors created it specifically to avoid patent infringement so it is widely used without fear.

The basic algorithm itself is based on the classic Lempel-Ziv 77 algorithm. It maintains a hash table filled with the recent strings and searches it to see if a particular string occurred before. If a string has a previous occurrence in the file, it is replaced with a tag containing a distance (less than 32k) and a length of the string (258 character maximum).

The algorithm uses two Huffman trees to add additional compression. One is used to handle the literal strings present on the first occurrence as well as the lengths of the strings. Another tree is used to compress the distance, and this helps optimize patterns.

The hash table is limited to three characters at the start of the string. This points to a list of all strings that start with those three characters. A string to be compressed is compared against all of the entries in the list to find the best (i.e., longest) match. Requiring three characters removes the problem that the algorithm will inflate the size of the file, because each tag takes three bytes to store.

The newest entries are stored at the front of the hash lists to ensure that the closest and thus smallest previous entries are found. This concentrates the distances on smaller numbers and makes it possible for the separate Huffman tree to get some good compression of these values.

In general, the algorithm is run on blocks of data that are as long as practical. The Huffman trees are transmitted at the beginning of each block, so it makes little sense to use small blocks unless transmission problems require it.

3.6 Conclusions

The dictionary-based compression mechanisms are some of the best known and most widely used programs around today. They do a good job with text and a passable job with binary object code. Image files are better handled by other methods like JPEG or JBIG.

Some patents on ideas from this Chapter are found beginning on page 183.

Dictionary coding programs gain significant speed advantages with better data structures. Anyone implementing a system like this should plan the structure for finding similar patterns in the dictionary. This is the most time-consuming part of the compression process, and a smoothly implemented data structure for storing the dictionary can lead to significantly faster compression ratios.

One interesting question is to compare the dictionary-based systems to statistics-based systems like Huffman coding or Arithmetic coding. Dictionary systems will outperform Huffman systems whenever fairly long words or phrases are used again and again in file. Political reporting, for instance, is filled with words like "Clinton", "impeachment", and "scandal" at the time this book is being written. A dictionary scheme would work well in this situation. It might also shine in places like the telephone book where there are many instances from a small selection of first names and a more widely varied, but still limited, selection of last names. The dictionary would quickly be filled with the different first and last names and the compressed file would only specify the combinations.

Dictionary schemes, however, would fare worse in situations where there are no long, repeated words and phrases. The stock tables, for instance, are filled with thousands of English words, but there is little of the repetition found in the telephone book. Many companies have synthetic words with the same letter probabilities as regular English text. In these cases, statistics-based systems will outshine dictionary systems.

No system is best for all files, and many commercially available systems test both dictionary- and statistics-based schemes on the file and end up choosing the one with superior compression.

Chapter 4

Arithmetic Compression

Chapter 2 describes how to use a statistical model of the data to create variable-length codes for a file. The most common patterns or words get short codes, and the others get longer codes. The Huffman coding algorithms described in that chapter make a good introduction to the process, but they are not optimal.

Arithmetic coding is a similar class of algorithms that can squeeze the data even more. The basic approach is very similar, but it finds the extra space to compress between the characters. It essentially packs the compressed codes for the characters in even tighter than before.

The basic conceit of arithmetic coding is that the entire file is compressed by converting it into a single binary fraction between 0 and 1. Every programmer knows that the package of bits known as a computer file can always be sliced and partition into any wide variety of numbers or letters in a number of formats. A single binary fraction may seem a bit extreme, because this implies that a file that is compressed to be 100,000 bits long is essentially a number with 100,000 bits of precision.

IBM pioneered arithmetic coding, publishing some of the earliest papers and also patenting some approaches. See the work of J. Rissanen and Glen G. Langdon, Jr. [RL79, RL81, Lan84].

Framing the process as a quest for a very precise number makes it seem daunting. In practice, the process is not much different from Huffman coding: a letter is read from the file, a statistical table is consulted, some computations are done, and then a variable number of bits for each letter code is added to the file. The biggest difference occurs in the step where some computations are done. Huffman codes algorithms build a tree and look for a path from root to leaf. This algorithm uses basic arithmetic.

A binary fraction is just like a decimal one. An n-bit value is just the integer representation of the number divided by 2^n. So ".01" is one-quarter, ".0101" is five-sixteenths.

One subtle difference responsible for the success of the algorithm is the fact that the codes for each letter blur together. In a Huffman-encoded file, each character is responsible for a set of bits, and each bit is carrying information about only one character. In an arithmetically compressed file, some characters work together to define some bits, and some bits carry information about adjacent characters. There may be some bits that only define one character, but there will be others that do more. This blurring is where the arithmetic algorithm gets the ability to outperform Huffman coding. In many cases, the algorithm doesn't need an exact number of bits to represent a character, but the Huffman algorithm requires an exact integer number. So it gets rounded up. These rounding errors can add a small but significant amount to the size of a file. Sometimes arithmetic approaches can outperform Huffman solutions by 10%.

See Figures 2.3, 2.4, and 2.5 for examples of how Huffman compression fails.

The basic algorithm begins with an alphabet, A, filled with characters or perhaps tokens created by aggregating several characters. Each token, x, occurs with some probability $\rho(x)$ defined by examing a file and determining how often a character exists. This probability model may be either a static one that is computed once at the beginning of the compression process or a dynamic one that adapts to the file. The model must somehow be shipped to anyone decompressing the file, and this is a major part of the overhead.

Huffman coding uses these probabilities to build a tree. Arithmetic coding uses these values to break up the number line between 0 and 1 into intervals. Let $x_1, x_2, \ldots, x_n$ be an enumeration of the tokens in A. Then the interval $0 \leq y < \rho(x_1)$ stands for x_1, $\rho(x_1) \leq y < \rho(x_1) + \rho(x_2)$ stands for x_2 etc. In general:

$$\sum_1^{i-1} \rho(x_i) \leq y < \sum_1^{i} \rho(x_i).$$

A token is compressed by choosing any y in the interval to stand for the value.

The key insight is that any value in the interval can stand as a compressed version of the token. This gives the algorithm some flexibility to find the best possible choice and perhaps squeeze out some extra space. Often, the best choice is the value with the shortest representation. Imagine that A consists of two characters with $\rho(x_1) = \frac{3}{5}$ and $\rho(x_2) = \frac{2}{5}$. Choosing $y = \frac{1}{2} = .1$ is the shortest binary fraction in the range $0 \leq y < \rho(x_1)$. $y = \frac{3}{4} = .11$ is the shortest binary fraction in the range $\rho(x_1) \leq y \rho(x_1) + \rho(x_2)$.

Choosing numbers like this will produce the same variable-length codes as the Huffman or the Shannon-Fano algorithm if the intervals are arranged in the same way. The same algorithms used to build the trees can be used to order the intervals in the same way.

Choosing one short binary fraction for each token, however, will never do better than Huffman or Shannon-Fano coding. The critical insight in arithmetic

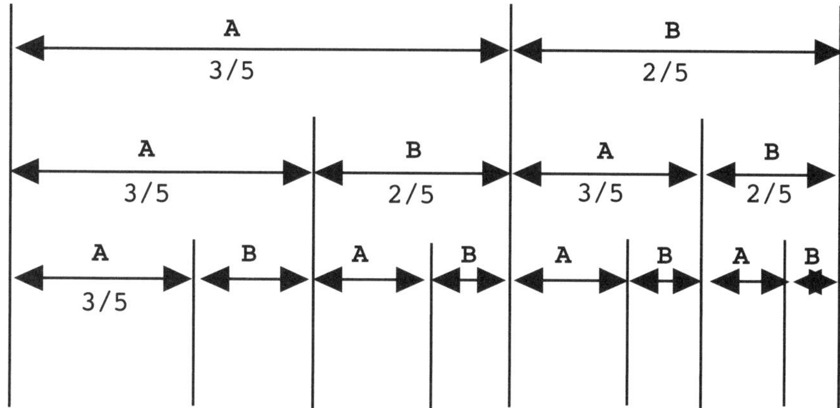

Figure 4.1: The number line between 0 and 1 split three times recursively. Each level encodes one token by delinating the intervals in the line corresponding to each character.

coding is that the interval-splitting algorithm can be repeated again and again. After it has been split numerous times, then the shortest number inside the one tiny interval can stand for a surrogate for all of the intervals. It is, in essence, applied recursively.

Figure 4.1 shows the number line between 0 and 1 after being split three times using the hypothetical example of a two-character alphabet where one character occurs 60% of the time and other occurs 40% of the time. The string 'AAA' corresponds to the interval between $0 \leq y < (\frac{3}{5})^3 = .216$, and the string 'ABB' corresponds to the interval $\frac{3}{5} + \frac{2}{5} \times \frac{3}{5} \leq y < \frac{3}{5}$.

The algorithm procedes by subdividing and subdividing each region into an upper and a lower boundary. One number inside this tiny sliver can represent all of the characters. Decompression can be accomplished by testing this one value against recursively subdivided regions. For instance, let the value $\frac{1}{8} = .001$ stand for the string 'AAA' using the intervals defined by Figure 4.1. The first character can be decoded as 'A', because $0 \leq \frac{1}{8} < \frac{3}{5}$. The second character is also 'A', because $0 \leq \frac{1}{8} < \frac{9}{25}$. The final character is also 'A', because $0 \leq \frac{1}{8} < \frac{27}{125}$.

4.1 Three examples

This section provides three examples. The first one uses characters where the probabilities of occuring are negative powers of two. This means that Huffman

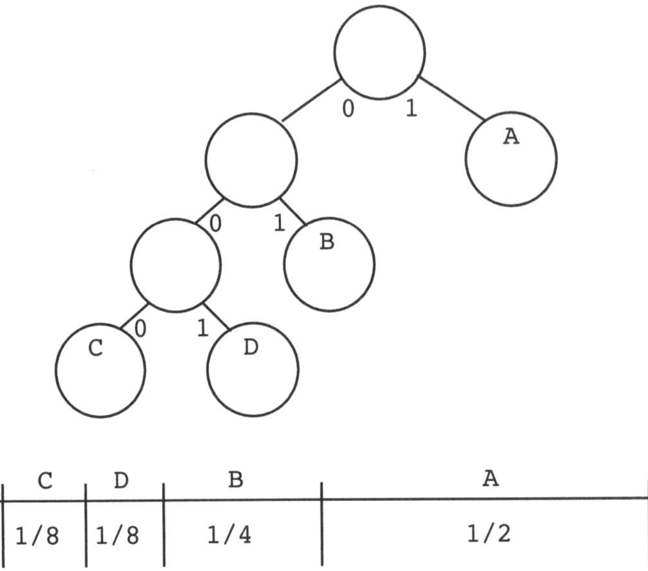

Figure 4.2: A Huffman tree and an arithmetic coding segmentation of the number line that are both designed to compress a simple four character alphabet where the letters occur with probabilities $\{\frac{1}{2}, \frac{1}{4}, \frac{1}{8}, \frac{1}{8}\}$.

coding can provide optimal compression for this file. The example will show how arithmetic compression creates the same output in order to illustrate the similarities between the algorithms.

The second and third examples will use probabilities that are not powers of two. This is much more likely to happen in reality, and these example will show how arithmetic compression can squeezes the extra space out of the file. It will illuminate the differences between the two approaches.

Begin with four characters, $\{A, B, C, D\}$, with corresponding probabilities $\{\frac{1}{2}, \frac{1}{4}, \frac{1}{8}, \frac{1}{8}\}$. Figure 4.2 shows a Huffman tree used to compress the file alongside an arithmetic coding interval. The segments of the interval are rearranged to make the arithmetic coding correspond to the Huffman codes.

Here's the result of compressing the string 'ABCDC'. Each step shows how the interval is reduced. The Huffman codes are provided for comparison.

Letter	Huffman Code	Lower Bound Fraction	Upper Bound Fraction
A	1	$\frac{1}{2}$	1
B	01	$\frac{1}{2}+\frac{1}{2}\times\frac{1}{4}$	$\frac{\frac{1}{2}+1}{2\times\frac{1}{2}}$
C	000	$\frac{1}{2}+\frac{1}{2}\times\frac{1}{4}$	$\frac{1}{2}+\frac{1}{2}\times\frac{1}{4}+\frac{1}{2}\times\frac{1}{4}\times\frac{1}{8}$
D	001	$\frac{1}{2}+\frac{1}{2}\times\frac{1}{4}+$	$\frac{1}{2}+\frac{1}{2}\times\frac{1}{4}+$
		$\frac{1}{2}\times\frac{1}{4}\times\frac{1}{8}\times\frac{1}{8}$	$\frac{1}{2}\times\frac{1}{4}\times\frac{1}{8}\times\frac{1}{4}$
C	000	$\frac{1}{2}+\frac{1}{2}\times\frac{1}{4}+$	$\frac{1}{2}+\frac{1}{2}\times\frac{1}{4}+$
		$\frac{1}{2}\times\frac{1}{4}\times\frac{1}{8}\times\frac{1}{8}$	$\frac{1}{2}\times\frac{1}{4}\times\frac{1}{8}\times\frac{1}{8}+$
			$\frac{1}{2}\times\frac{1}{4}\times\frac{1}{8}\times\frac{1}{8}\times\frac{1}{8}$

This table shows the same calculations in binary:

Letter	Huffman Code	Lower Bound Binary	Upper Bound Binary
A	1	$.1_2$	1.0_2
B	01	$.101_2$	$.11_2$
C	000	$.101000_2$	$.101001_2$
D	001	$.101000001_2$	$.10100001_2$
C	000	$.101000001000_2$	$.101000001001_2$

This table makes it clear how the lower bound of the interval follows the Huffman codes for this tree. The final lower bound is just a binary fraction with the same pattern of zeros and ones as dictated by the Huffman coding.

In the second example, the four letters in the alphabet occur with probabilities that are not negative powers of two. For this example, let $\{A, B, C, D\}$ occur with probabilities $\{.7, .12, .1, .08\}$. Figure 4.3 shows the Huffman tree and arithmetic coding interval for this example as well. The Huffman tree is the same as the one in Figure 4.2, despite the fact that the probabilities are significantly different. The tree is still the best possible choice for these probabilities.

Here's a table showing the compression of the string 'AAAB'. The binary representations of the fractions are left out to save space.

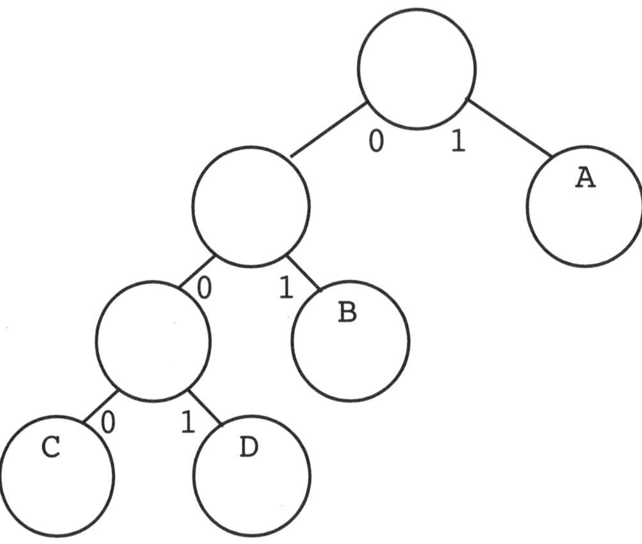

Figure 4.3: A Huffman tree and an arithmetic coding segmentation of the number line that are both designed to compress a simple four-character alphabet where the letters occur with probabilities $\{.7, .12, .1, .08\}$. This diagram is very similar to Figure 4.2, because both sets of probabilities generate the same Huffman tree. The intervals, however, are significantly different.

Letter	Huffman Code	Lower Bound Base 10	Upper Bound Base 10
A	1	$.3$	1
A	1	$.3 + .3 \times .7 = .51$	1
A	1	$.3 + .3 \times .7 + .3 \times .7 +$ $.3 \times .7 \times .7 = .657$	1
B	01	$.3 + .3 \times .7 + .3 \times .7 +$ $.3 \times .7 \times .7 +$ $.18 \times .7 \times .7 \times .7 = .71874$	$.3 + .3 \times .7 + .3 \times .7 +$ $.3 \times .7 \times .7 +$ $.3 \times .7 \times .7 \times .7 = .7599$

The Huffman tree will compress this string to be '11101'. The arithmetic compression reduces it to some number in between $.71874 = 1011011111111111_2$ and $.7599 = 1100001010001_2$. The shortest binary fraction in this interval is .11. A simple way to find the shortest binary fraction is to start at the left of both binary expansions and scan to the right looking for the first different bit. Keep everything that is the same, and then add a 1 at the end.

The arithmetic compression algorithm has reduced this string of four characters to two bits — a substantial win that is less than one bit per letter. This example, however, produced such a stunning win by a bit of luck. If the fraction $.75 = .11_2$ weren't in the interval, then it would probably have taken about four digits to represent a number in the interval.

The third example illustrates how arithmetic compression can occasionally do worse than Huffman compression. In the second example, the string 'AAAB' came pretty close to matching the statistical profile used to do the compression $\{.7, .12, .10, .08\}$. In this next example, the string, 'ABCDC', will be compressed with the same model. This string, however, will emphasize some characters that are supposed to appear rarely. The result will take more bits than Huffman compression.

Here's a table showing the arithmetic compression of the string 'ABCDC'. The Huffman codes are provided for comparison. The binary strings are left out of the table to save space.

Letter	Huffman Code	Lower Bound Decimal	Upper Bound Decimal
A	1	$.3$	1
B	01	$.3 + .18 \times .7 = .426$	$.3 + .3 \times .7$
C	000	$.3 + .18 \times .7+$ $.7 \times .12 \times .08 = .43272$	$.3 + .18 \times .7+$ $.7 \times .12 \times .18 = .44112$
D	001	$.3 + .18 \times .7+$ $.7 \times .12 \times .08 = .43272$	$1.3 + .18 \times .7+$ $.7 \times .12 \times .08+$ $.7 \times .12 \times .1 \times .08$ $= .433392$
C	000	$.3 + .18 \times .7+$ $.7 \times .12 \times .08+$ $.7 \times .12 \times .10 \times .08 \times .08$ $= .43277376$	$.3 + .18 \times .7+$ $.7 \times .12 \times .08+$ $.7 \times .12 \times .10 \times .08 \times .18$ $= .43284096$

The final lower and upper bounds in binary are

$$.43277376 = 0110111011001010010000101111_2$$

and

$$.43284096 = 0110111011001110101010101001_2.$$

The only step left is to find some value inside this range. The best choice is the value with the shortest binary value: $.01101110110011_2$. This takes 14 bits — two more than Huffman compression.

Most uses of arithmetic compression will not be as extreme as any of these examples. When larger files are compressed, the patterns of letters will do a better job matching the statistical profile used to guide the compression. There should rarely be cases where the arithmetic compression underperforms Huffman compression.

Entropy is discussed on page 22 in Chapter 2.

In most cases, arithmetic compression will just come closer to achieving the theoretical minimum predicted by the entropy of the stream. In the example described in Figure 4.3, the entropy of the four-character stream is $.7log_2\frac{1}{.7} + .12log_2\frac{1}{.12} + .10log_2\frac{1}{.10} + .08log_2\frac{1}{.08} = 1.35097$. The expected performance of the Huffman compression on this dataset is $.7 + .12 \times 2 + .10 \times 3 + .08 \times 3 = 1.48$. This means that the arithmetic compression should do better that 1.48 bits per character but not more than 1.35 bits per character.

One way to see this easily is to look at the most extreme case of a two-character alphabet: 0 and 1. Imagine that 0 occurs 75% of the time in a truly random distribution. Huffman compression can only build a simple tree that represents each character with another single bit; call them 0 and 1 for simplicity sake.

Arithmetic compression, on the other hand, can squeeze out all of the entropy. The result should be about 81% of the original. As one character begins to dominate, the size of the arithmetically encoded file will shrink, while the Huffman file will remain the same.

4.2 Programming Arithmetic Coding

The algorithm for arithmetic coding can be set forth fairly simply. Let `width[i]` be $\rho(x_i)$. Let `offset[i]` be an array set to $\sum_0^{i-1} \rho(x_i)$, the sum of the widths of all the characters smaller than it. These values are used to narrow the two variables `lower` and `upper` into a thinner and thinner slice of the number line. Let the variable `slice` keep track of the width of this slice.

Start with `lower` set to 0, `upper` set to 1, and `slice` set to 1. Here are the steps for the inner loop.

1. Get next token or character. Let `i` be the index of it.
2. `lower=lower + slice * offset[i]`
3. `upper=lower + slice * (offset[i]+ width[i])`
4. `slice=slice* width[i]`

This can be repeated until the final answer is found.

Many programmers first assume that creating an arithmetic coder will require doing some fractional mathematics with millions of bits of precision. Files are often this long even after compression, and the algorithm clearly searches for one long binary fraction.

Any implementation can benefit from the fact that the variables `lower` and `upper` change by smaller and smaller amounts on each pass through the loop. As `slice` gets smaller and smaller, some bits in these bands become fixed. When they are fixed, they don't need to be computed anymore. Plus, they can be output because they are also part of the answer. The shortest binary fraction between `lower` and `upper` can be found by keeping all of the bits that are the same and adding a 1 at the end.

The simplest solution to keeping these values in range is to test the first bit of `lower` and `upper`. If the bit is the same, output the bit and then multiply the values of `lower`, `upper`, and `slice` by 2. If `lower` or `upper` goes above 1, subtract one. This multiplication does not destroy the precision or the accuracy of the answer — it merely removes the fixed bits from consideration. This is because the first n bits of `slice` are zero whenever the first n bits of `lower` and `upper` agree. It is the equivalent of a bitwise left shift.

If this multiplication is done after every pass through the algorithm, enough precision should be kept in a double precision floating point variable.

The decompression algorithm is equally easy. The variables `lower`, `upper`, and `slice` are still used to maintain information. But, in this case, the data from the compressed file must be compared against the possible values from `offset[i]` until the right match is found.

Here's a summary of the steps. Initialize `lower` to zero, `upper` to one, and `slice` to one. Let `answer` be the compressed data. It can be loaded bit by bit as necessary. It shouldn't be stored as a floating point number, but as some fixed-point value that approximates this.

1. Set `i` to be 0.
2. Increment `i` until `lower+slice * offset[i]` is greater than `answer`.
3. Output x_i, which is the character to be decoded.
4. `lower=lower + slice * offset[i]`
5. `upper=lower + slice * (offset[i]+ width[i])` [1]
6. `slice=slice* width[i]`

This algorithm can also discard bits once `upper` and `lower` agree upon them. Just keep discarding bits by multiplying, `answer`, `lower`, `upper`, and `slice` by two.

Precision and Arithmetic Coding

The three examples starting on page 51 used fixed fractions to do the calculations for the arithmetic coding. They used perfect accuracy to represent the values. This was largely possible because the examples were short. While it is always possible to used fixed-point arithmetic and infinite precision arithmetic packages to compute the ranges, the extra work is usually not worth the trouble. Storing the values of `lower`, `upper`, and `slice` as double precision, floating-point values can suffice.

[1] A better solution is to use `offset[i+1]`, which is equal to `offset[i]+width[i]`. This isn't as easy to understand at first.

The effects of this approximation are difficult to measure but usually easy to ignore. The value of `slice`, for instance, will slowly under-approximate the true value, because the double precision floating point number will only keep the first 50 or significant bits of the value. This truncation may end up being significant if a huge file is compressed.

Imagine that n bits of `lower`, `upper`, and `slice` are kept current. This means that there is an error term associated with each value that is smaller than 2^{-n}. Call this ϵ. At each step, `slice` is multiplied by the width of some interval, `width[i]`, which is a fraction. This means that ϵ is still less than 2^{-n}. This value is added into `lower` and `upper`, and the amount of ϵ will not go in. This will only be a problem after a significant number of steps. If $n = 50$, then there will not be a problem with most files, which are generally shorter than 2^{32} bits.

Another area where imprecision arises is in the statistical model of the characters or tokens in a file. The values of $\rho(x_i)$ are gathered from a real file, and these may not have accurate representations as binary fractions. The value of $\frac{1}{3}$ is a repeating binary fraction .010101 .... These will also have error terms that could prove significant if the file is large enough.

One simple defense is to approximate all of the fractions with n bits and then ensure that all of the values of $\rho(x_i)$ add up to 1. This simple step will reduce the errors.

4.3 Products Using Arithmetic Coding

Arithmetic coding algorithms are becoming more common in products, but it is still not part of the most popular versions. The *hap* files are one of the standards found on the Internet that use the algorithm.

Timothy C. Bell, John G. Cleary, and Ian H. Witten are authors of a popular piece of freely available source code written in C. It can be found at

`ftp://ftp.cpsc.ucalgary.ca/pub/projects/ar.cod/`

This was described in their book [BG90]. Another version of the software was written by Fred Wheeler to be more object-oriented and easily adapted to compressing multiple streams. It can be found at

`http://ipl.rpi.edu/wheeler/ac/`

This version will adapt its statistical model as it passes through the data.

4.4 Conclusion

The process of arithmetic coding is, at the very least, intellectually interesting. It illuminates some of the problems with Huffman compression and shows how some careful arithmetic can pack the data even tighter.

The algorithm can be more difficult to code than the dictionary coding schemes from Chapter 3 or the Huffman schemes from Chapter 2. The precision arithmetic is not sophisticated mathematics, but it requires more attention to detail than simple data structures.

This chapter has not examined some of the different ways that characters can be aggregated into tokens. Many of the same schemes from Chapter 2 are just as applicable here. If aggregating two characters into a token will reduce the overall entropy estimate of the file, then it will also reduce the final compressed size of the file. Arithmetic compression does a good job of approximating the entropy model.

This chapter has also avoided examining the ways that more advanced statistics can be used to improve compression. The same schemes that work in Chapters 2 and 5 will work with this model as well. If you want to spend more time creating bigger models of the statistical profile of the data stream, then you can also create bigger, more complicated sets of intervals to do the compression. The more complicated the statistics, the better the compression — until the overhead from transmitting these statistics overtakes the compression.

The JPEG image compression algorithm is described in Chapter 10.

Arithmetic compression is also used as a component in other compression schemes. The JPEG algorithm, for instance, can use either Huffman coding or arithmetic coding to compress the coefficents.

Readers should be aware that IBM and several other companies have received patents for various versions of arithmetic compression. This may limit the usefulness for people who don't want to pay royalties.

Chapter 5

Adaptive Compression

The algorithms in Chapters 2 and 4 have one thing in common. They create a static model of the data in the file and use this to compress the entire file. This approach is perfectly valid, but it has a few shortcomings. The greatest may be the fact that files often have patterns that are very local. The phone book, for instance, has a few pages where the name "Smith" is common, and many pages where it is relatively rare. Researchers addressing compression have noticed this effect and tried to capitalize upon it when building algorithms that adapt to the current data.

Dictionary algorithms described in Chapter 3 are good example of general adaptability. Most will build up a copy of the dictionary as they work through the document, essentially adapting the dictionary to the words in the file as each new word is processed. Some will even confine the definition of the dictionary to a particular window of the file. That is, they will only keep a definition of a word around for n characters. If it occurs again outside of that window, it will be treated like a new word.

Adaptive techniques can also reduce the amount of overhead in a file. Static methods will usually ship some table summarizing the statistical information at the beginning of a file. Adaptive methods can reduce this amount, in some cases, because they often don't ship any additional data. The compressing and decompressing functions build up the same tables simultaneously as they adapt to the data. The effects of adapting to the data are often mitigated by the fact that extra characters must be included to signal the construction of the table.

Huffman coding, described on page 16, is a good example of an algorithm that ships it as overhead.

Most statistical approaches can be changed, or adapted, to be adaptive through one simple idea. Instead of building a statistical model for all of the characters or tags in a file, the model is constructed for a small group and updated frequently.

Here's a list of different ways the model can be updated:

- The model can be constructed for the first i characters in the file. After each character, the model is updated, and the result is used to compress the next character. This may be the simplest adaptive approach around.
- The model is maintained to keep statistical information on a sliding window of n characters. After each new character is processed, the model is updated to include this character and remove the one n characters before. This approach does a good job of capturing local effects, but it also requires storing n characters in addition to maintaining all of the tables.
- A new model is computed for blocks of m characters. This approach can be combined with one of the first two by restarting the adaptive model building every m characters. This approach works well when there are many rarely occurring characters. Restarting the model construction removes them from the model, which allows the remaining ones to be compressed a bit more.

The biggest problem with adapting the model to fit the data on a character-by-character basis is complexity. The software is harder to write, and it takes longer to run. It is not the best solution for data that must be compressed and decompressed on the fly. It is better for archives and situations where the data is only unpacked occasionally.

This complexity can be reduced to some extent by carefully designing the algorithms so that the adaption can be accomplished quickly. Well-designed data structures can make this fairly simple.

5.1 Escape Codes

Adaptive algorithms have one additional problem. They must delineate between new data and compressed data. Static algorithms do not have this problem because they create a file with compressed data and overhead. Adaptive algorithms don't have the overhead. They mix the new data into the file for the algorithm to discover and adapt to.

This is a fairly abstract way of saying adaptive algorithms use escape codes that signal that the next block of bits should be treated as raw, uncompressed data. That is, this is a new character that isn't in the model yet. The escape code is usually inserted in the model as a character itself.

This escape character is essentially the overhead. It occurs once for each new character not found in the model. The code is usually smaller than one entry in

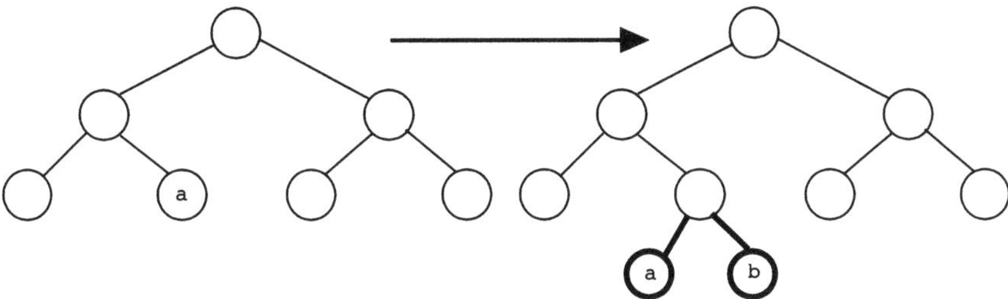

Figure 5.1: The process of adding a new node to a Huffman tree involves finding the node with the lowest occurrence in the file. A new node is created to handle both nodes. Here the node containing the character 'a' is replaced by an interior node that joins the nodes for 'a' and 'b'.

the table transmitted with a static model, so it usually means that there is less overhead.

Most of the dictionary algorithms in Chapter 3 use escape codes to indicate new entries in the dictionary.

5.2 Adaptive Huffman Coding

Huffman coding requires building a tree that describes the statistical makeup of a file. Adaptive Huffman coding modifies the tree as each character is processed. The tree manipulations are fairly straightforward and easy to code because they can be expressed recursively. Not all of the tree needs to be changed or adapted at each step.

The tree is built of nodes, $\{n_1, \ldots, n_k\}$. The value $v(n_i)$ reflects the number of times that the character associated with that node occurs in the file. When a character is processed, the tree is adapted to reflect its existence. The steps look something like this:

- Determine whether the new character is part of the tree. If it is already in the tree, then add one to its count, $v(n_i)$.
- If it is not in the tree, find the leaf node with the lowest count. Create a new node for the character, and add it to the tree at this place by creating a new interior node that joins the two nodes. Figure 5.1 illustrates this procedure.
- In either case, the tree must now be rebalanced. In most cases, the Huffman tree is still in good shape. The tree is still the optimal way to compress the

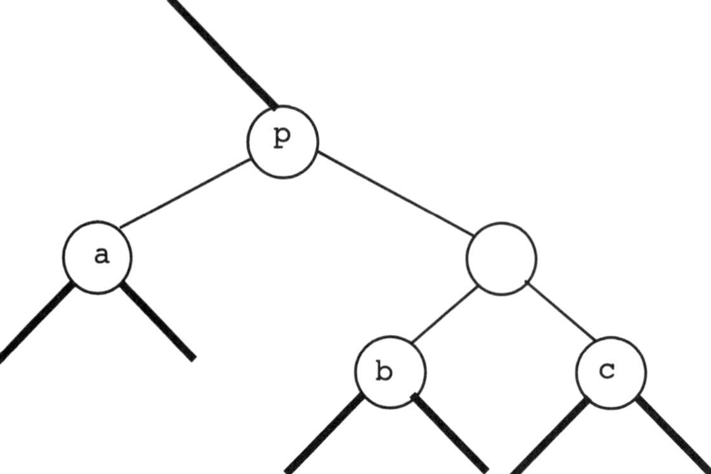

Figure 5.2: The tree is unbalanced if $v(b) > v(a)$ or $v(c) > v(a)$. If this happens, n_b or n_c should be swapped with n_a and the values of nodes adjusted accordingly.

characters represented by it. In some cases, the tree will become imbalanced because the counts are wrong.

- Work up the tree incrementing the values of $v(n_i)$ for all interior nodes up to the root.
- Figure 5.2 shows a generic situation found in the tree that may indicate that the tree is unbalanced. This happens when either $v(n_b) > v(n_a)$ or $v(n_c) > v(n_a)$. If this is the case, then n_b or n_c should be swapped with n_a. Note that n_a, n_b, and n_c may not be leaf nodes. They could be interior nodes with descendants of their own.
- The swapping algorithm should be done repeatedly until the tree is balanced. The algorithm should begin with the node of the character being processed and place it in the position of either n_b or n_c. If a swap takes place, then the step should be repeated for the parent node, n_p. The process is repeated until no swapping is necessary.

The swapping process is pretty straightforward and can be implemented by a recursive algorithm. Processing each character takes only a few additional steps to maintain the tree.

If an adaptive windowing solution is used, then characters must also be deleted from the tree. This process is essentially the reverse of the process of addition. The steps look like this:

- Locate the character. Decrement the count, $v(n_i)$.
- If the count goes to zero, remove the node. This is the opposite of the process described in Figure 5.1.
- Now, rebalance the tree. The tree is unbalanced if $v(n_a) < v(n_b)$ or $v(n_a) < v(n_c)$, as illustrated in Figure 5.2. Swapping nodes fixes this.
- The swapping process begins at the root and works its way down the tree following the path from the root to the node representing the character being removed from the model.

Some patents on ideas from this chapter are found beginning on page 209.

While these adaptive Huffman encoding algorithms look impressive, they often don't add much extra compression. This happens in situations where the statistical makeup of the data is fairly homogeneous, as with English-language text. The distribution of characters is fairly even except in situations where large numbers of names distort the process. Still, even large numbers of words that begin with 'z', for instance, still come with the same major vowels and consonants. The overall statistics don't change too much, and this often means that the overhead of adaptive Huffman encoding isn't worth the effort.

5.3 Windows of Data

Another common solution used with all of the algorithms in this book is limiting the compression function to a smaller part of the file. Reducing the scope of the function to a smaller subset of the data allows it to customize itself to that section. Data that vary widely between parts of the file can be a big advantage. It can be a waste, however, if the file is relatively homogeneous, because each segment usually needs some overhead.

Focusing the function on a narrow window can also increase the stability of the data. Even losing one bit in a file compressed with Huffman encoding can jeopardize everything that comes after that missing bit. Losing a character can mess up the dictionary algorithms that encode a dictionary entry as an offset from the current location. Restarting the compression every n bytes limits the damage to one n byte packet.

Narrow windows are also useful in situations where the file is too large to be used in one piece. The video compression functions like MPEG achieve most of their compression by transmitting only the parts of the image that change between frames. Still, the systems include a complete image every n frames to help recover from errors, assist editing, and prevent gradual degradation of the image.

5.4 Conclusion

Many of the algorithms described in the book do some adaptation to the local data by changing the statistical model as they move through the file. In most cases, this adds a small amount of additional compression, but in a few cases it can make a big difference. An obvious example might be the phone book, which is alphabetically sorted. The letter 'z' is much more common at the end.

Many of the algorithms use a windowing algorithm to focus the statistics. This is usually pretty inexact, because it is often defined by a hard-coded number. Many dictionary algorithms look at the last 2^i characters for a match, for instance.

Unfortunately, many of the published algorithms do not try too hard to globally optimize the adaption process. This might make it possible to find hard boundaries for regions in the file that have particular statistical behavior. A sliding window doesn't do a good job matching the 26 different parts of the phone book.

Chapter 6

Grammar Compression

One of the more recent developments in compression technology is the use of higher-order grammars to to produce excellent compression of structured files such as text. Some of the latest experimental results suggest that the approach is one of the best known solutions that can routinely generate results less than two bits per word if the file is long enough. The catch is that the process is highly computationally intensive, and it can require a large amount of memory. Also, the greatest savings do not come until a large block of text is processed.

Craig G. Nevill-Manning and Ian H. Witten suggest that better compression results can be gained by interpreting the sentence structure of English text [NMW97, NMW98].

The process is similar to the dictionary systems described in Chapter 3, but it adds an additional layer of complexity. The algorithm scans through the text looking for patterns, and ,when it finds them, it adds them to a table just as the dictionary system does. The difference is that it includes some additional logic for recognizing more sophisticated patterns. Ideally, these more sophisticated approaches can generate even better savings because they can describe the data with more sophisticated and thus smaller patterns. Clearly, equating "more sophisticated" with smaller is not a hard-and-fast rule, but one that must be worked toward with careful programming and adaptation.

The most common form of representation is the *grammar*. This is a logical structure that is fairly similar to the natural languages used by humans. In fact, the structure was first invented by linguists like Noam Chomsky who were looking for ways to describe human communication in a concise and consistent format. It was later adopted by computer scientists who needed a way to represent computer languages.

In this case, the algorithms attempt to build a grammar that will represent the data. A grammar is a collection of *variables* and *terminals*. The terminals are the letters, characters, words, or tags that are part of the final output. In some cases, these algorithms work on a character-by-character basis, and the terminals are just the 128 printable ASCII characters. In other cases, the algorithms work on a word-by-word basis, and the collection of words forms the collection of terminals.

In the system by Nevill-Manning and Witten [NMW97, NMW98], adjectives, adverbs, verbs, and other phrases are split apart and compressed with separate tables. The results can yield compressed files as small as about 1.7 bits per character at a cost of long compression times and taxing loads on memory. The system requires tables about three times the size of the file being compressed to hold the grammar rules inferred from the text.

Variables are internal symbols used by the grammar to produce the final output. They are essentially placeholders that will eventually be converted into terminals in a process known as *production*. There is always one special variable known as **Start** that begins the entire process. The variables will always be shown in **boldface**, while the terminals will be described with regular type.

Here's an example of a grammar that also illustrates the connection with natural language:

Start	→	**noun verb**
noun	→	Mary ‖ Bob
verb	→	went to dinner **where** ‖ went dancing **where**
where	→	in **direction** Ohio. ‖ in **direction** New York.
direction	→	northern ‖ southern

The variables are listed on the left hand side of the arrow (→), and the *productions* are shown on the right. In this case, there are multiple productions on the right hand side separated by a double vertical line (‖). This means that both productions are equally valid.

A grammar, G, is said to be a collection of variables, terminals, and productions. Usually a special variable for starting the entire process is designated. Each grammar defines a collection of strings that can be produced by the grammar, usually represented as the language $L(G)$. The string "Mary went to dinner in northern New York" is in the language described above, but the string "Mary ate ground steak in New Jersey." is not. Here's an illustration of how the **Start** variable leads to a string in the language:

Start
noun verb
Bob **verb**
Bob went dancing **where**
Bob went dancing in **direction** Ohio.
Bob went dancing in southern Ohio.

There are a wide variety of different grammars in use. One of the most common forms is called the *context-free grammar*, which gets its name from the fact that there are no restrictions on how or when the variables can be expanded by productions. The counterpart to this is the *context-sensitive grammar*, which allows certain productions to be used only in certain places. The grammar above is context free, but it would be context sensitive if **direction** could not be expanded into the terminal "northern" after **noun** was expanded into "Bob." Context-free grammars are studied fairly exhaustively in computer science, and most computer languages can be described in this form.

John Hopcroft and Jeffery Ullman describe grammars in depth in their book[HU79]. They provide good theoretical insight into which patterns can be described with grammars and which can't. One tool, known as the pumping lemma, *makes it relatively easy to classify some patterns.*

Grammars are usually good ways to define fairly complicated patterns. For instance, the pattern 'abbabaababbababaabbabaab' can be represented with the following grammar:

Start	→	**CCC**
A	→	ab
B	→	ba
C	→	AABB

In fact, a dictionary like the ones used in Chapter 3 might be considered as a *regular grammar*, a simpler form of the grammar where there is only one variable on the right-hand side. These grammars accept the class of *regular languages*, a more restrictive class of languages.

The hope is that the grammar-based program will be able to thread together more complicated productions with multiple variables that will allow large sections of the repetitive text to be represented with just a few variables.

6.1 SEQUITUR

The SEQUITUR algorithm developed by Craig Nevill-Manning, Ian Witten, and D. L. Maulsby is a good example of a fairly simple approach that generates good results with text. The authors report results of around two bits per character on files larger than a megabyte or so. [NMW97, NMW98]

The compression algorithm works by watching for repeated pairs of either terminals or nonterminals. The terminals are letters from the alphabet, and the nonterminals are the variables introduced as new rules or productions are created. The algorithm is fairly simple. It seeks to ensure that "no pair of adjacent symbols appears more than once in a grammar" and "every rule is used more than once."

These rules are converted into a fairly simple algorithm:

1. Let **Start** be a nil string. There are no productions defined yet.
2. For each character in the file, add it to the end of **Start**, and repeatedly ask the following questions until they both answer no:
   (a) Are the last two symbols at the end of **Start** found on the right-hand side of some production? The word *symbols* means either terminals or nonterminals. If the two can be found, then replace the two of them with variable that produces that pair.
   (b) Can the last two symbols be found in some other place in the **Start** string? If so, create a new variable, and replace the repeated pair with this variable. Add a new production that turns the variable into the pair.
3. Remove rules that are only used once in the grammar by compressing them. That is, if **A** →**BC** , **B** →de, and **C** → af, and the productions for the variables **B** and **C** are only used once, then replace them with one rule, **A** →deaf.

Here's a table showing the algorithm operate:

New Letter	**Start**	New Production
a	a	*none*
b	ab	*none*
b	abb	*none*
a	abba	*none*
b	**A**b**A**	**A**→ab
a	**A**b**A**a	*none*
b	**BB**	**B**→**A**b
b	**BB**b	*none*
a	**BB**ba	*none*
b	**BB**b**A**	*none*
b	**BB**b**B**	*none*
a	**BB**b**B**a	*none*
b	**BB**b**BA**	*none*
b	**BB**b**BB**	*none*
	Cb**C**	**C**→**BB**

In the end, the production **A**→ab is only used once, so it is rolled into the production for the variable B: **B**→abb.

Nevill-Manning and Witten show that the algorithm should take an amount of time that is linearly proportional to the size of the incoming text, despite the fact that a fairly large tree can be constructed. They base this proof on the assumption that both the **Start** string and the table filled with productions can be stored as hash tables that offer constant access time on average. If this is the case, which it often is for reasonably sized texts, then the process of adding new letters to the end of **Start** ends up taking a constant time on average.

This algorithm can be fairly easy to implement. The production building steps can be accomplished with a single array of integer pairs. The integers between 0 and 127 should be assigned to the ASCII letters, which serve as terminals, and the integers greater than 128 represent variables. An additional integer can be included to keep track of the number of times the rule is used. The pairs can be stored in a hash table to speed up lookup.

In the end, a Huffman-like algorithm can be used to assign bit strings to the variables to save more space.

6.2 Code Compression

Grammar-like systems are popular in compressing software packages, because the sequences of operations used by computers is often fairly complicated but composed of common patterns. For instance, a common set of instructions will be: Load address A from memory, load address B from memory, add the two values, and finally store the result in address C. These sequences are common in RISC operations.

The common approach is to separate the addresses A, B, and C from the stream, because they vary from instance to instance, and find a way to compress the sequence of operations: load, load, add, store. If the operations alone are compressed, the results can be fairly dramatic. Many of the operations take up an entire instruction word, but they are often found in long repetitive patterns. Basic routines like `for` loops are often converted into the same basic string of operands, and they can all be compressed into one short idiom. [1]

[1]Code compression is a well-studied region of both compression research and microprocessor architecture. Some of the more studied papers include ones by Jens Ernst, Christopher W. Fraser, William Evans, Steven Lucco and Todd A. Proebsting [EFE+97, FMW84, FW86, FP95, Fra99] ; Charles Lefurgy, Peter Bird, I-Cheng Chen, and Trevor Mudge [LBCM97, LM98, BM96, CBM96, Che97]; Vincent Cate and Thomas Gross [CG91], Mauricio Breternitz Jr. and Roger Smith; [JS97] ,Michael

After the addresses are stripped away, any of the compression techniques in this book can be used. The dictionary approaches are often the best, because several strings of opcodes are much more common than the others. The best approach is to develop a fixed dictionary for each particular processor in advance. It is possible to analyze the code samples, lock in a standard set of choices, and then use Huffman-like coding to assign bit strings to the entries.

Grammar-based methods can be even better because there are usually several sets of operands that are combined in different ways. For instance, the basic increment command `i++` is common in C because it is used in loops and standalone instances. A grammar-based system can assign one code for the set of opcodes that implement the instruction when it is standing alone and add it into the larger pattern used for the loop.

[LBCM97] reports that it is able to compression PowerPC binary code by 39% and i386 code by 26%.

Code compression algorithms, however, are often limited by the fact that they must be implemented in silicon. While the compression can be arbitrarily complex because it is only done once during compilation, the decompression must be done on the fly by the chip. If it is too complex to decompress the bit stream, the chip will either require too many circuits, be too slow, or both.

These limitations are not always present. Java byte code, for instance, is an important and widely used technology on the Internet. It is usually interpretted with a software virtual machine and occassionally cross-compiled into native code for the processor. In either case, the instruction codes would be decompressed by a software layer. Any compression algorithm used for Java code would not need to worry about excess transistors, but it should still be relatively fast. At this time, Java uses a version of PKZIP to bundle its source code, but this is not optimized for code.

Abstract Syntax Graphs

The best solution may be to transmit the abstract syntax graphs of the instructions. Normally, compilers analyze the syntax trees and interleave the instructions in order to provide optimal use of the processor's time. If data will take several clock cycles to fetch, then other work can be done instead of holding up the processor. This means that the patterns are often interleaved, which adds more complexity and reduces compression. Another solution is to leave the code in the graph so it can be compressed more effectively.

Kozuch and Andrew Wolfe [KW94, KW95]; Stan Liao, Srinivas Devadas, Kurt Keutzer, Steven Tjiang, and Albert Wang [LDK+96, Lia96, LDK+96, LDK+95b, LDK+95c, LDK95a]; M. Franz and T. Kistler [FK96, Fra96, Fra94]; Alex Aiken and Alex Nicolau [Aik88, AN91]; and A. Wolfe and A. Chanin [WC92]

Trees and graphs can be compressed by numbering the nodes and keeping a list of all of the arcs that join the nodes. General graphs can be stored as a list of arcs while trees can take even less space if their structure is exploited.

The nodes in these graphs are numbered, and the number is all that really matters. It can be "compressed" by using a flexible length coding scheme for the values. That is, the short numbers take up a short number of bits. A simple scheme is to use groups of i bits as digits. These packets of i bits encode 2^i different values, but one of them must be set aside as the stop or termination value.

For example, if $i = 3$, then let 111 be the termination criteria. The rest of the number is stored in base $2^i - 1 = 7$. So the value 3 becomes 011111, the value 7 becomes 001000111, and the value 13 becomes 001110111. Shorter values of i are good if there are many small numbers like 0, 1, or 2, and longer ones work better for greater distributions.

The chief advantage of this scheme is that it does not have a maximum limit. Arbitrarily large numbers can be encoded for an overhead of i bits plus the small added cost of recasting numbers into base $2^i - 1$. If there is a known limit on the size of the values and this known limit is small (like 16), then fixed blocks of bits can be used.

Using these schemes, a graph can be encoded as a list of numbers where each pair encodes an arc. An extra bit (or two) can be added if the graph is directed.

Trees can be compressed even more. If each node has at most j descendants, then nodes can represented as j bits, where 1 signifies the existence of a descendent and 0 signifies that none exists.

The nodes themselves are arranged in depth-first order, which can be calculated with this function:

```
depthfirst(node){
  if (left!=null) {
    ans="1";
  } else {
    ans="0";
   }
  if (right!=null) {
    ans=concat(ans,"1");
  } else {
    ans=concat(ans,"0");
   }
  print ans;
  if (left!=null) {
```

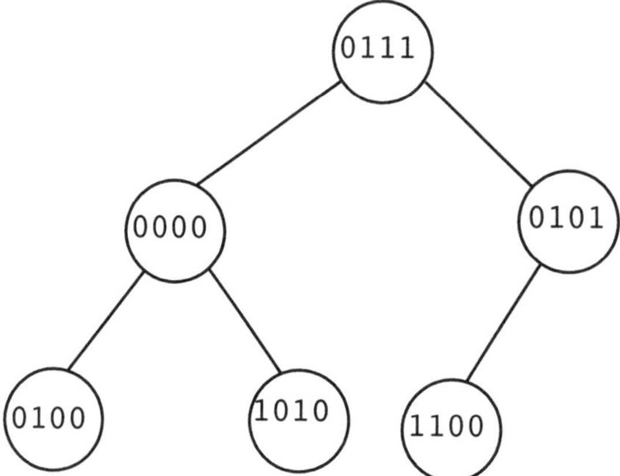

Figure 6.1: A tree that is encoded using two bits to encode the nodes in depth-first order. This is 111100001000.

```
    depthfirst(left);
  }
  if (right!=null) {
    depthfirst(right);
  }
}
```

In this case, `depthfirst` is a recursive function. `node` is the current node, which comes with two descendants, `left` and `right`. (This can be extended to arbitrarily large sets of descendants.) The function `print` stores the two bits of representation in the right place.

Figure 6.1 shows a simple tree that can be encoded by using the depth-first rule. 111100001000 is the result of executing the function on the tree in the figure and starting at the root. The data in the nodes can either be inserted into the string of data or appended at the end. 110111110000000100001010100101001100 is the result. The spaces are added to improve readability.

If the information inside the nodes is added after, then each of the techniques from the book can be applied to encoding the string of bits that describe the tree. Clearly, run-length encoding has great potential for larger trees. Statistical methods can also do a good job with full trees.

6.3 Conclusion

Grammar-based compression has great potential to shrink files, because it can recognize more complicated patterns than the dictionary approaches described in Chapter 3. The research is still beginning at this writing, and it is possible that more research will develop more sophisticated approaches for identifying patterns that can be represented in short grammars.

Code compression is also a natural application for grammar-based approaches, because most computer code is generated from languages like C, and these computer languages are all based on context-free grammars. These grammar-based approaches can be integrated with complete representations of the abstract syntax graph if necessary.

Chapter 7

Programmatic Solutions

One of the more fundamental concepts of computer science is the strange mutability of a program and data. Most software comes with a fixed wall that divides the information on the disk into either instructions for the processor or data to be manipulated. An address book comes with the instructions for sorting the names, displaying them on the screen and printing them on paper. The addresses are separate. Instructions are active. Bits are passive.

Most good programming classes quickly teach that this wall is not fixed, but movable. It's very easy to add more intelligence to the instructions so that the data can be sloppier and perhaps shorter. Basic compression routines are a good example. Many programs add the ability to decompression basic files into the program, and this allows them to save space on the data. It's easy to take the opposite route. Some programs are deliberately simple to save on execution time. This often requires larger data files that are better structured.

Some software designers have taken this insight to an extreme by designing a data structure that is, for many intents and purposes, a programming language. That is, the data can include all of the basic features of programming, such as loops, variables, subroutines, and conditional jumps. This can often make the data file quite short because it provides a good way to add structure to the file.

One of the best and most common examples of this approach is the PostScript file format that is one of the most prevalent standards in the printing and graphics design world. This book was converted into PostScript at one stage in its production, and many of the printers recognize the format.

PostScript was developed by Adobe systems and based, in part, on ideas outlined with the Interpress format developed at Xerox PARC.

PostScript includes many of the instructions that are pretty obvious. There are instructions that say "place the letter 'A' at position $(100, 200)$ on the page" or "draw a line between $(101, 304)$ and $(101, 205)$." These commands make up the bulk of the language, and they're used frequently.

But PostScript also comes with other commands for making loops, adding if–then conditionals, or even creating subroutines. These generic features are powerful enough to allow programmers to create endless combinations of instructions for filling the pages with complex patterns of lines and dots. Most word processing companies, for instance, have written a set of common subroutines that are included as a header in all files sent to a PostScript printer. They include basic kerning routines and other page layout features.

These basic subroutines reduce the size of the overall file. The subroutines are called to do all the details of laying out the header and footer of each page. This can save a fair amount of space depending upon the complexity of the subroutines and the ingenuity of the programmers.

This chapter will describe several basic approaches for adding programmability to data files in order to compress files. The content might better be described as programming language theory, because it deals with some of the design decisions that others have made in creating their file formats.

7.1 PostScript

PostScript is a great example of a simple language that includes many of the good ideas from programming language theory without becoming too large or unwieldy. The basic structure might be described as a cross between the syntax of Forth and the flexibility of Lisp. Both are well known in the programming world for being efficient and powerful languages that can describe complicated algorithms in a minimal number of keystrokes. Combining the two is a good way to begin compressing data.

The basic syntax in PostScript uses a stack and postfix notation. That is, the data come first and the operand comes at the end. So adding two and two is represented as `2 2 add`.

Hewlett-Packard continues to sell calculators that use this postfix structure, which they call Reverse Polish Notation.

The PostScript instructions are interpreted by reading through the data and placing the individual tokens on the stack. When one of the tokens turns out to be an operand, then it is executed using the data that are on the stack. In the case of the sequence `2 2 add`, the first two digits are pushed on to the stack. The plus operand then pulls the two digits off, adds them together, and places the result, `4`, on the top of the stack.

All of this pushing and popping may seem to be overly complicated, especially when the problem is just adding two digits. But the structure pays off, because more complicated equations can be represented with a short sequence. Consider $(10+3) \times (5+7)$. This becomes `10 3 add 5 7 add mul` in this postfix notation. A postfix interpreter is also quite easy to write so it takes up little memory.

The Lisp-like features of PostScript make it possible to manipulate data and then convert them into a program. Subroutines are created in PostScript with a command called `def` that binds an array of commands with a name. The syntax is demonstrated in this example from the PostScript "Red Book" [ano86]:

```
/average {add 2 div} def
```

In this case, a forward slash ("/") is added as a prefix to the word "average", and this signifies that it should be treated as a literal. It is pushed onto the stack without being interpreted.

The curly brackets ("{" and "}") denote the beginning and end of an array of commands. This is how arrays are added into PostScript. They are pushed onto the stack as one entity.

The command `def` will bind the word "average" with the array containing the three commands. This is stuck in a big table that matches words with arrays. Once a command is defined, it can be invoked by including it in a string of instructions. The sequence `10 30 average` will leave the result `20` on the stack.

PostScript also comes with an if–then construct that allows branching. It looks like this:

```
first second gt
{if true } {otherwise this } ifelse
```

There are six boolean operators, `eq, ne, gt, ge, le,` and `lt`, that can be used in this expression. If necessary, they can be combined with the usual `and, or, xor,` and `not`.

There are also the basic loop constructs defined for the stack. A `for` loop looks like this:

```
1 2 10 {doMe} for
```

The first value, `1`, is initial value of the loop counter. The second value, `2`, is the increment. The third, `10`, is the final value, and the process ends when the value passes it. The array of instructions makes up the body of the loop.

Execution with the stack is a bit strange. There are four steps to each iteration of the loop:

1. The counter is tested to see if it is larger than the final value. If it is, then the loop stops immediately.

2. The value of the counter is pushed onto the stack.

3. The instructions in the loop block are executed. They will normally consume the counter value pushed on the stack, but, if they don't, the value stays there.

4. The counter is incremented.

This basic `for` loop can be used in many different ways. Executing `10 10 100 {} for` will put the values `10 20 30 40 50 60 70 80 90 100` on the stack, because the blank loop block will do nothing with the values.

PostScript also comes with the `forall` instruction, which is applied to an array. It pushes the values from the array onto the stack at each iteration instead of using the counter. The command `loop` will execute a block until the block executes the `exit` command.

The subroutine, the loop, and the conditional statement are the three basic components of a programming language. Everything else flows from these constructs. PostScript is a pretty good example of how to implement these constructs without heavy overhead. Languages like C or Java may be a bit easier to understand and use, but they also require more complicated parsers and compilers. PostScript originally ran on the Apple Laserwriter, which came with a 68000 and a small amount of memory. (It was often more powerful than the Macintosh sending it commands.)

PostScript also comes with another fairly simple construct that can be very powerful. The `/name save def` command will take the current state of all objects and bind it to the name "name". This is an ideal way to pause before executing a subroutine. The command `name restore` brings it back. There is also a subset of this known as `gsave` and `grestore` that will bind up the graphics state (color, origin, line width, and so on). While these are not necessary to write complicated code, they make it easier to shift state, a task that is important when doing jobs like laying out pages.

PostScript Specific Instructions

PostScript also comes with a number of features that are tuned to specifying documents, that is, commands for putting letters and lines in particular places. These commands are optimized for creating documents. While you may not need these

Figure 7.1: A simple, fast PostScript illustration.

in your application, you can take some inspiration in the way that they were implemented.

The cleverest part of PostScript is the way that the scale and the location of the origin lines can be quickly changed. Executing the command `1.2 .8 scale` will immediately begin multiplying all x coordinates by `1.2` and all y coordinates by `.8`. This can be extended by including a complete projection matrix for skewing the drawing.

This command can be combined with the `translate` and `rotate` command to create great patterns in a small amount of room. The `x y translate` command will immediately move the origin by `x` and `y` units. The `theta rotate` command will immediately rotate the coordinate axes by `theta` degrees.

Here's a simple example:

```
1 1 5
{1 0 translate
  30 rotate
  drawme
  pop}
for
```

This `for` loop will execute the command `drawme` five times in five different locations that are evenly spaced along a curving line. At each step, the origin is moved over one unit, and then the axes are rotated. Whatever actions the `drawme` routine does are repeated at each spot.

Figure 7.1 shows the result of running these commands through a PostScript intepreter:

```
%!PS-Adobe-2.0 EPSF-1.2
```

```
%%BoundingBox: 0 0 281 182
/Times-Bold findfont 48 scalefont setfont
100 20 translate
10 -1 1
   { 0 0 moveto
      .1 mul
      setgray
      10 rotate
      (PostScript) show
      } for
```

The loop steps down from 10 to 1. At each step, the loop counter is placed on the stack, and then the instructions between the curly brackets are executed. In this case, the first instruction (`0 0 moveto`) doesn't use the loop counter. It pushes two zeros on to the stack and then moves to that location. The second instruction (`.1 mul setgray`) uses the loop counter and multiplies it by ten before using it to set the gray level being drawn. Finally, the coordinate axis is rotated by 10 degrees with each step. This rotation builds upon itself with each step.

While it is possible to make fairly complicated patterns by combining the routines with loops and strange recursive procedures, the `scale`, `translate`, and `rotate` are usually used to place figures or text in certain locations. The figures in this book are created independently in PostScript and then scaled, translated, and inserted into the right place by the TEX language. These are excellent commands for the page layout world.

PostScript also comes with a full complement of rules for specifying lines, Bezier curves, and arcs. There are also rules for colors, shaded fills, and images. Fonts are just defined by Bezier curves and linked with particular characters. While these features are an essential part of PostScript, they are not particular inspiring. The structure is not hard to imitate or modify.

7.2 Conclusions

Compressing data with executable files is a great approach that can be quite useful in many situations where the data are well structured. This is often the case when the data are generated by some program and simply saved before being used by another program. The structure developed from the data can just be exploited.

This technique isn't going to be much use with acquired data, such as the numbers found in images or audio files. There is enough random noise to make it hard to find deep, repeating patterns that can be coded with a PostScript-like scheme. The techniques described in Chapter 9 probably come as close to this as possible by fitting repeating wavelets to the data.

Of course it is possible to take this to extremes. Figure 13.1 shows a fractal drawing that was actually computed by the PostScript interpreter. The points were calculated by the printer, not the computer I used to prepare this manuscript. This was quite inefficient and shows how the programmatic features can lead to solutions that may not be the best for everyone. In this case, a short program draws tens of thousands of points on the page. It might be the best solution if the image needs to be sent over a very expensive channel where every byte is precious. In most everyday cases, however, it is silly to force the printer's low-end computer with little memory to do the work.

It is also important to realize that these features can backfire if they're used indiscriminately. Many word processors create a standard set of page lay out routines that are included as a header for every file sent to the printer. One might kern a line of text in a particular way. Another might lay out the page numbers. This makes great sense, but it can actually lead to large files that are swollen with the overhead of these functions. It is not uncommon to watch a word processor take a file with two words ("hello world") and produce a PostScript file that is over 10k bytes long. Almost all of this is PostScript functions that are never used.

Chapter 8

Quantization

Most of the compression algorithms at the beginning of the book are aimed at compressing data from abstract collections of symbols, letters, or tokens from some set, Σ. These algorithms, like Huffman coding (Chapter 2) or Lempel-Ziv (Chapter 3), look for patterns in the data that can be replicated with a shorthand. When the data are decompressed, the exact patterns are reconstructed and the new data are exactly the same. This *lossless* compression is often the only solution for the abstract collections of symbols.

This chapter, on the other hand, will deal with numbers. Numbers are abstract collections of symbols that come with a special function that defines how close they are to each other. This function allows people to create compression algorithms that come *close* to replicating the data. The phrase "almost only counts in horseshoes and hand grenades" is not correct. It also counts in this arena where the process known as *quantization* lets people compress a stream of data and *almost* reproduce it when it is decompressed.

This chapter will deal only with values from the real numbers and the integers. More abstract algebras are left to the reader.

The process should not be new to anyone. Numbers already require a certain amount of quantization to handle in regular life. The value of $\frac{1}{3}$, for instance, can't be described with base ten: .33333 . . .. When the repeating fraction is truncated to say .3333, the real number $\frac{1}{3}$ is replaced with a number that is shorter (four digits instead of infinity) and close enough for most purposes. Rounding off values is a classic technique that is a simple version of quantization.

The algorithms in this chapter take the process a bit further and provide a number of different ways to quantize values that are all more flexible and adaptive. They're just fancy ways to round off the numbers to save space.

The biggest difference in these algorithms is how much overhead they take.

All of these systems need to provide a convenient understanding for how numbers are represented so both the compression function and the decompression function will operate in the same way. In a sense, these algorithms aren't about finding a way to compress data. They're really about finding a good system of representing numbers that produces a small data file.

Naturally, there are some trade-offs in the process. The fraction $\frac{1}{3}$ does a perfect job of representing the value with only two characters. The repeating decimal value .3333 is just an approximation, and it takes four characters. While fractions can do a good job of representing natural numbers with more precision, they can be awkward to use. Comparing values takes several arithmetic operations, and the basic arithmetic operations take several steps. Plus, the values grow larger and larger with several operations producing large values: $\frac{1}{3}+\frac{1}{8}=\frac{11}{24}$. For all of these reasons, fractions aren't a great scheme for representing numbers.

The overhead determines how adaptive the schemes may be. Some use fixed schemes for rounding off the values and converting them into the right "quantum level". Others change the levels to fit the data. While this can provide more compression, it requires shipping along the new quantum levels to the decompression algorithm. This overhead, like all overhead, will cut into the amount of compression.

Some of the algorithms presented here are used to quantize vectors. These are just multidimensional numbers, so all of the techniques from quantizing real numbers should apply. In practice, the algebra can get more complicated, so simple algorithms are often substituted. These simpler algorithms often have other features that are desirable.

In general, most of the algorithms in this book are mainly used for image compression. This is largely because these are the most common data that do not need to be reproduced exactly. It makes little difference if the picture of the face comes back with a extra amount of red in the cheeks as long as this extra amount is small enough to go unnoticed. Image files are also large enough to make any approximation seem desirable even if it comes at the cost of some inexact reproductions. There is no reason, however, why it can't be applied to general numerical data.

8.1 Basic Quantization

The basic approach to quantization replaces each real number, x, with an integer, i, such that x and the value $mi+b$ are as close as possible. Both m and b are constants set up in advance to do the best job conforming to the data. The value of

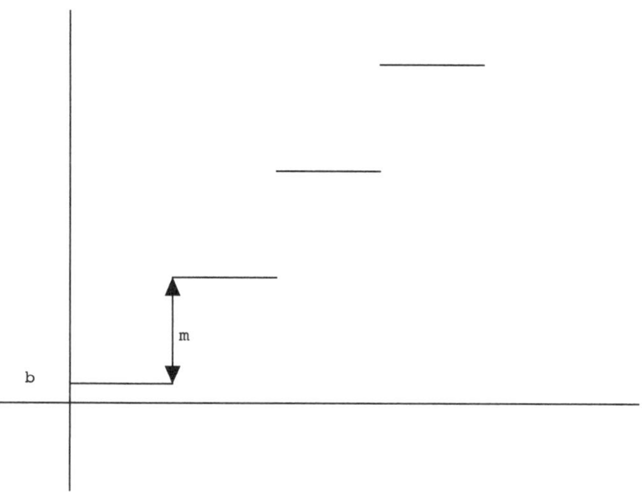

Figure 8.1: The range between x_{min} and x_{max} is replaced with quantized levels.

m is the size of each *quanta*. Larger quanta provide greater compression at the cost of lower precision. Smaller quanta provide better accuracy but less compression.

The amount of compression depends upon the number of quanta needed to cover the numbers in the data sample. If the file consists of values between x_{max} and x_{min}, then the range can be broken up into k quanta of size, $m = \frac{x_{max}-x_{min}}{k}$. The value of b is set to be $\frac{x_1+m}{2}$. The file is compressed by converting it into numbers between 0 and $k-1$. The amount of compression depends upon how many bits it takes to represent the integers between 0 and $k-1$. This is normally $\log_2 k$, but it may be less if additional schemes such as Huffman compression are used to represent these values. Often, some quanta are more common than others, and Huffman compression can squeeze even more space out of the data stream.

Huffman compression is combined with quantization in some image compression schemes, such as JPEG. See Chapter 10.

Decompression consists of replacing the quanta with some plausible data value. A simple solution is the midpoint of the range between $mi + b$ and $m(i+1) + b$. Figure 8.1 illustrates this point.

There is no reason why the size and position of the quanta need to be arranged in a linear fashion. Using a linear arrangement makes sense for collections of data that are uniformly distributed across the range between x_{min} and x_{max}, but not all sets of data fit this criteria. Some data sets may lend themselves to to schemes based upon the log of the value. In this case, i should be chosen so that $mi + b$ is as close as possible to $\log x$. This is useful for data sets such as wealth, where the order of magnitude is more important than the actual value.

In general, any function describing the distribution of data can be converted into a scheme for constructing quanta. A good example is the familiar normal distribution, $\rho(x) = e^{-x^2}$, that says the probability of a value occurring depends upon its distance from zero. The goal is to slice up this range into quanta so that each quantum is equally likely to occur.

This can be accomplished with integration. In this case,

$$\int_{-\infty}^{\infty} e^{-x^2} = \sqrt{\pi},$$

so the range can be split up into k segments by finding $x_0 \ldots x_k$ so that

$$\int_{x_i}^{x_{i+1}} e^{-x^2} dx = \frac{\sqrt{\pi}}{k}.$$

In this case, $x_0 = -\infty$ and $x_k = \infty$, but finite sets will have finite endpoints. Using an odd value of k ensures that the middle version of x_i will be set at zero. Once these endpoints are calculated, they can be used to compress the file by converting data values into the index of the quantum in which they fall. The easiest way to calculate these endpoints is through numerical integration of the function, a result that is often known as the error function in engineering.

Decompressing the data requires coming up with a new replacement value for all of the data elements that fall within a quantum unit. This is where the error is introduced, because one value is chosen as a surrogate for many. There is no best choice for all cases, but a simple choice for the range x_i to x_{i+1} is d_i, where

$$\int_{x_i}^{d_i} e^{-x^2} dx = \frac{1}{2k}.$$

Figure 8.2 illustrates the function being sliced up into $k = 8$ regions. The smallest quanta are found near zero where the values are the most common. The largest ones occupy the tails stretching out into infinity.

Evaluating the Error

Compressing a file by quantizing it and decompressing it will introduce a certain amount of error into the file because the decompressed file will not have the same values. It is possible to estimate the error in advance with some mathematics.

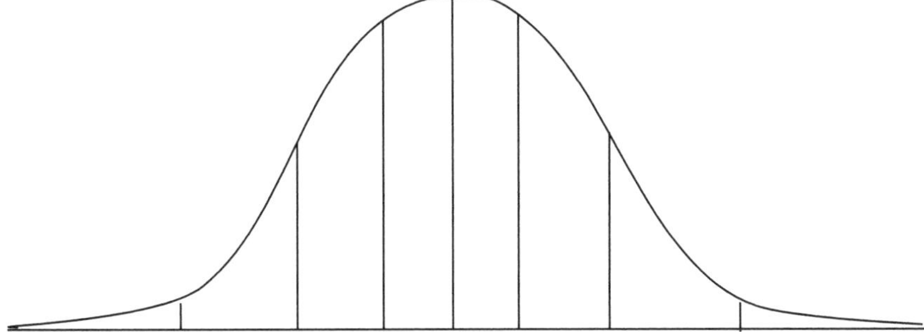

Figure 8.2: The normal distribution is sliced up into $k = 8$ ranges.

If the data are equally distributed across the range between x_{min} and x_{max}, then the median error will be $\frac{m}{4}$. Half of the values will be closer than $\frac{m}{4}$, and half will be farther away. None will be more than $\frac{m}{2}$ units away.

If the logarithms of the data are equally distributed, then the logs will exhibit the same behavior. The errors in the reconstructed file, however, will vary from quantum unit to quantum unit. The largest units will have the largest errors.

8.2 Adaptive Quantization

The last section described how to split up the real numbers into a section of k quanta defined by $k + 1$ points $x_0, x_1, \ldots, x_{k-1}, x_k$. The positions could be constructed by some real valued function, $\rho(x)$, describing how the data is distributed.

In many cases, the data does not fit some pattern with any consistency. It may be log distributed in one file, evenly distributed in another, and clustered in strange places in a third file. While static quanta will still work for all three cases, adjusting the size and position of the quanta can produce less error during reconstruction.

The previous section expressed the model of the data as a continuous probability distribution. This made it easy to construct fairly complicated quanta by integrating the probability distribution. This section will describe the data as a finite set of n numbers: $\{a_1, \ldots, a_n\}$. This a closer approximation to the description provided many compression functions. The functions are usually handed a data file and asked to compress it without any more sophisticated knowledge about the distribution of the data. The only solution is to construct the model of the data from the data itself.

The simplest solution is to scan the data file and calculate x_{min} and x_{max}.

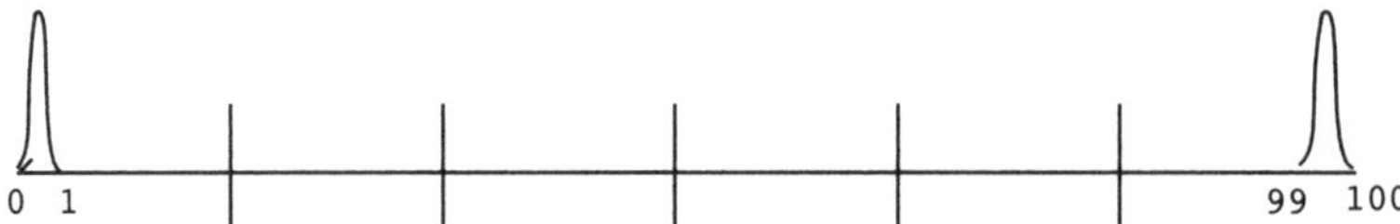

Figure 8.3: The probability distribution of the values in a hypothetical file where the data is clustered in two ranges between either 0 and 1 or 99 and 100. Clearly, the basic approach of splitting the range into k quanta fails, because most of the quanta are unused. In this example, $k = 6$.

The values of m and b can be calculated after the number of quanta, k, is decided upon. This approach uses a small amount of overhead, k, m and b, and produces reasonable solutions.

A set of better approaches tries to minimize the error by looking for clusters of data and arranging the endpoints of the quanta around the positions of these clusters. This can reduce the amount of error when reconstructing the file.

Here's a simple example illustrated in Figure 8.3. A file consists of two clusters of values. Half are found between 0 and 1, and the other half are found between 99 and 100. If the basic static algorithm described on page 86 were used, then the quanta would be constructed from $x_{min} = 0$ and $x_{max} = 100$. If $k = 2$, then two quanta of size $m = 50$ would be used. The compression process would reduce the values to one bit apiece ($i = 0$ or $i = 1$), but the decompression process would replace them with the midpoints of the quanta, $d_0 = 25$ and $d_1 = 75$.

The error would be terrible, because the data is not evenly distributed. The error for each term would vary between 24 and 25. The solution also fails if k is larger than two. As the illustration in Figure 8.3 shows, all but two of the quanta are unused. Clearly, a better solution can be found.

Greedy Algorithm for Quanta Location

This solution is not particularly efficient, and it may be occasionally problematic, but it is easy to understand and guaranteed to reduce the error. The approach works by choosing the "midpoints" of the quanta, $D = \{d_0 \ldots d_{k-1}\}$, first. The endpoints, $x_0 \ldots x_k$, are chosen to be in between them.

Here's the algorithm. Begin by setting d_1 equal to one value of a_i. This is usually chosen at random. Using a more sophisticated approach will do a better job, but the random selection will usually do a pretty good job. Repeat these steps until there are k "midpoints".

1. Do a test compression of the file by assigning each data value a_i to a quantum by finding the closest d_j to it.

2. Find the a_i with the biggest error. That is, with the greatest distance between it and its d_j.

3. Add a_i to the set of midpoints.

When the algorithm is finished, the "midpoints" are sorted and renumbered. The endpoints, $x_0 \ldots x_k$ can be calculated to be halfway in between: $x_0 = x_{min}$, $x_k = x_{max}$, and $x_i = \frac{d_{i+1} - d_i}{2}$.

This algorithm is good for reducing the maximum error. Each step is devoted to this process. It just looks for the worst error and reduces it to zero. The algorithm is not particularly efficient, because it requires rescanning the entire data set for each step.

Some use the nomenclature, $O(kn)$ to mean that the algorithm takes an amount of time that is related to the product of k and n. This is usually defined to mean that there exists a constant c such that ckn is greater than the amount of time used to compress any file with n units by finding k quanta.

The algorithm begins to fail, however, at reducing the average error when there are just a handful of points that lie outside of the range. To understand this, imagine the same pathological example of a file where most of the thousands of points lie between either 0 and 1 or 99 and 100. The algorithm fails, however, if there are just a few extra points at, say, 27, 37, 42, 50, 58, 65, 80, and 85. If $k = 10$, then the algorithm will end up with a set of midpoints like $D = \{0.74, 27, 37, 42, 50, 58, 65, 80, 85, 99.12\}$. While this will reduce the maximum error to less than 1, it will do little to reduce the average error much further.

If there are thousands of data points in the set, then it might be more desirable to ignore the outlying points and absorb the occasional maximum error by selecting midpoints such as $D = \{0.2, 0.4, 0.6, .8, 1.0, 99, 99.2, 99.4, 99.6, 99.8\}$. In this case, the average error will be well below .2.

Random Algorithm

Another choice is to simply choose k values of $\{a_1, \ldots, a_n\}$ at random to serve as D. This approach does a good job of matching the probability distribution. Big clusters of points are much more likely to have elements from the cluster chosen to serve in D. Small clusters and occasional points are likely to be ignored.

This version is likely to have good average luck, but it may have large errors. It is a bad choice if the maximum error is important.

One of the major advantages of this approach is simplicity. The algorithm requires a random number generator that makes k choices. This is pretty straightforward.

Subdivision Algorithm

One popular solution is to start with one big quantum unit that covers the entire range between x_{min} and x_{max}. This unit can be subdivided until all of the midpoints, $x_1, \ldots, x_{n-1}$, are identified. The subdivision can take many forms, but one of the most popular solutions is to find the range with the greatest number of points and choose the median point as a new midpoint.

Choosing the median ensures that half of the points in a quantum unit end up in one of the new quanta, and the other half end up in the other. It effectively makes sure that each of the quanta represents the same amount of points.

This recursive subdivision works best when the final number of quanta, k, is a power of 2. This is not a real liability because using a quanta count that is a power of two makes it easier to pack the quantum value into a fixed number of bits. If $k = 2^i$, then i bits can be used to represent each compressed number as the number of a quantum unit.

Huffman coding is discussed in Chapter 2.

If k is not a power of two, than a Huffman coding scheme can be grafted on to the system for converting the quantum unit into a set of bits that will represent the number. In some cases, it makes little sense to subdivide a quantum unit any further. This is usually because it is small enough to guarantee to reduce error to below the desired threshold.

Figure 8.4 illustrates how this might work with an example drawn from a dataset that uniformly covers the range between x_{min} and x_{max}, which for the sake of illustration are 0 and 16. The first several rounds of subdivision produced four equal regions between 0 and 4, 4 and 8, 8 and 12, and 12 and 16. At this point, each of the regions has the same number of data points that will eventually be assigned to it. For some reason, only the region between 4 and 8 is subdivided any further, in this example into regions between 4 and 5, 5 and 6, and 6 and 8. In practice, this will be because the data is not evenly distributed over the range, and the other three ranges are small enough. This example is evenly distributed to make the example easier to follow.

The algorithm for assigning Huffman-like codes to the regions is easy to understand. Start the algorithm with a null string assigned to the region between x_{min} and x_{max}. Every time a region is subdivided, add a "0" to the string, and assign it to one half. Also, attach a "1" to it, and assign that to the other half.

The simplest algorithm for creating this subdivision is to sort all of the numbers in the file. The median values can easily be extracted by jumping to the right element in the sorted list. If this file is too large to sort, then a representative sample might be taken and sorted in its stead.

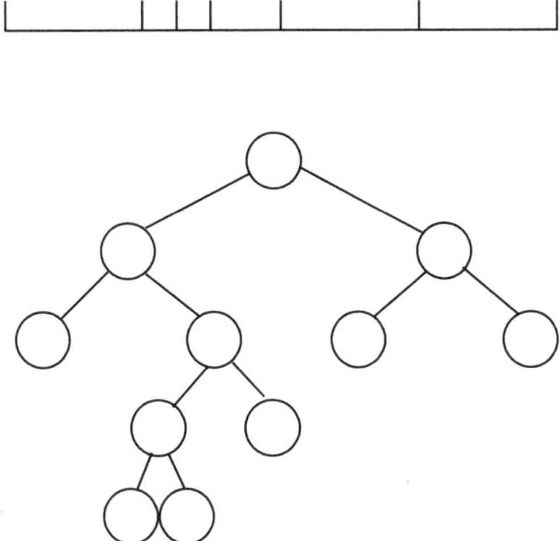

Figure 8.4: A region divided into six quanta with a Huffman tree drawn underneath. This shows how the Huffman addresses can be used to add a bit of extra compression. Some of the quanta are not subdivided as much as the others, and these can be represented with shorter bit strings.

8.3 Vector Quantization

In many applications, the data values aren't numbers but vectors of several numbers. The most common instance of this today is in color images where each pixel is represented by three numbers that indicated the amount of red, green, and blue in the color at that pixel. Color images are fairly large, so people often want to compress them. One of the simplest ways is to do *vector quantization* of the values. That is, find a smaller set of vectors that do a good job of representing the colors in the image and replace each pixel with its closest match from this smaller set.

The GIF (Graphics Interchange Format) is a popular format that offers this ability. A graphics artist can choose a pallet of 2^i different colors for $i = 1, \ldots, 8$, and each pixel will be represented by i bits. The format bundles the list of 2^i colors at the beginning of the file and follows it with a list of pixels represented by an number pointing to an index in the file.

Naturally, all of the vector quantization algorithms apply to other data types as well. They're just abstract ways of comparing vectors of numbers with a list and finding the closest match.

[CGV96] gives a survey of using quantization for subband coding.

Each of the one-dimensional algorithms described in the first part of this chapter can be used to quantize vectors as well. The biggest change is the method for estimating the distance between vectors. While this is fairly straightforward with one-dimensional numbers, it can be more complicated with vectors. The simplest metric for measuring the difference is the square root of the sum of the squares of the differences of each elements. That is, if a vector, v, is made up of n elements $[v_1, v_2, \ldots, v_n]$, then the distance between v and w is

$$\sqrt{\sum_1^n (v_i - w_i)^2}.$$

There are times, however, when some data terms are more important than others. In colors, for instance, the eye is more sensitive to green light than to either red or blue. So if humans are looking at the colors, then it makes sense to weight the values differently. This might be done by multiplying each term in the sum by a weighting value, α_i. This produces

$$\sqrt{\sum_1^n \alpha_i (v_i - w_i)^2}.$$

This section will assume that there is some distance function $\delta(v, w)$ that will return a real number that is the distance between two vectors v and w.

Greedy Algorithm for Vectors

The greedy algorithm for vectors is just like the greedy algorithm for scalars. The algorithm creates a set of vectors $D = \{d_1, \ldots, d_n\}$ one by one. At each step, each vector in the file is compared against each vector in D. The one that is farthest away from all of the vectors currently in D is added. To be more precise, if a_i are the vectors in the file, then the algorithm searches for i that produces

The greedy algorithm for scalars is described on page 90.

$$\max_i \min_j \delta(a_i, d_j).$$

This value is added to D, and the process continues.

The greedy algorithm has the same problem in vector space as it has in scalar space. The outlying points are quickly chosen as members of D even if a vast majority of the points are in clusters. This may be the ideal choice if you want to reduce maximum error, but it may not generate the lowest average error.

The section of Voronoi diagrams on page 97 gives some suggestions for speeding up this approach.

Subdivision Algorithms

The subdivision algorithms for vectors are also similar in spirit to the subdivision algorithms for scalar quantities. The biggest difference is that the process of subdividing multidimensional spaces is more complicated. It is also a bit more complicated to determine the "center" of each region that acts as the surrogate for all of the points when the file is decompressed.

The subdivision algorithms for scalars can be found on page 92.

Here's one algorithm for subdivision. It splits the region by geometric arguments instead of looking for the most populous. Of course, it is also possible to split the regions by their population. The algorithms described for scalar quantities beginning on page 92 still work here.

1. Start with one region that covers the entire space of vectors in the file to be compressed.

2. Repeat the following step until there are n regions left.

   (a) For the dimension of each region, find the minimum and maximum values of the vectors contained in that region.

This algorithm comes from Paul Heckbert [Hec82].

(b) For each dimension of each region, compute the "length" between the minimum and the maximum values. This may be done by simply subtracting, or it could be accomplished by using a more complicated distance function. For instance, some people use a measure of "luminosity" when this algorithm is used to compress images. The eye is more sensitive to green than red or blue, and this function measures these effects.

(c) Find the longest distance in all of the regions, and then cut this region. There are several ways to cut the region in "half". The simplest is to measure along the longest dimension and find the halfway point. This may accomplished with pure division if the dimensions are simply treated as geometric quantities. If a more complex distance metric is in effect, then some more complicated math may be in order.

Another way to split the region is to find the median of all of the data points in that region. This guarantees that that clump of data in the region will be split in half. This is a useful solution if the clump is off to the side of the region.

3. For each region, choose a representative vector that can be used to reconstruct the data at decompression.

The process of finding a representative vector for the region can be accomplished in a variety of ways. The simplest solution is to take the geometric center, which is defined by the midpoint along each dimension.

Another solution is to average the values in the region. This works well if the points are concentrated in one corner of the region.

In his analysis of the algorithm, Heckbert suggests that the structure of the U.S. Congress is a good metaphor for understanding the differences between this algorithm and algorithms that split the regions based upon count. This algorithm is based upon the size of the region, and it is similar to the U.S. Senate, which gives each state two seats. The House of Representatives, on the other hand, apportions the seats based upon the population. Regions with many people get many seats, while sparsely populated lands get few. Clumping algorithms based upon splitting up the regions are more analogous to the House of Representatives.

This difference can be important in image compression where a large error can be more jarring than a slightly larger average area. Imagine an image dominated by a face. Most of the pixels are flesh, but a few in the eyes are green. There may also be some deep ruby flecks, perhaps from some jewelry. If a population–based approach is used to split up the regions, then the flesh-toned pixels will dominate the splitting process. The few non-flesh-toned bits will end up lumped into several large regions, and there will be a large error produced when one single color is chose to represent all of the colors in the region. This is one way people end up with red eyes in digital photographs. This is one of the reasons that Heckbert's approach described above is popular for images. It ensures that none of the regions are too "big".

Voronoi Diagrams

One of the popular topics in computational geometry is computing the Voronoi diagram of a set of points. This construction takes a set of points, $D = \{d_1, \ldots, d_n\}$, and splits up the space into regions that are closest to each particular point. These diagrams can be used to help quantize a file filled with vectors by making it a bit easier to locate the closest point.

Let x_i stand for the region associated with the point d_i. This region is defined as the intersection of a number of half-spaces. In the case of one-dimensional quantization, a single point is able to split the line in half. In two dimensions, a line does the job. Let $\{h_{i,1}, h_{i,2}, \ldots, h_{i,m}\}$ stand for the m half-spaces that define region x_i.

The goal is to find an easy way to find the right region, x_i, that contains some vector a_j. Here's a straightforward approach:

1. Choose a representative set of k half-spaces from the collection of half-spaces that define all of the regions in the space.

2. For each region x_i, determine whether it is in or out of the k half-spaces.

3. This will produce a k-bit value that can be used as an index into a hash table.

This hash trick has limitations. Some regions are not completely inside a half-space. That is, they're cut in half by the plane used to define the half-space. This means they must be added to the hash table multiple times. In fact, if there are k half-spaces defining the regions, and l of them cut a region x_i, then x_i must be added to 2^l different entries in the hash table.

This approach can also be used to help compress data, even when the regions are constructed by another method.

The collection of points can be chosen either at random or by looking for vectors that represent all corners of the space well. The random selection is simple to do and will generally produce a good average error in most cases. Sophisticated selection methods do a better job avoiding maximum errors.

8.4 Dimension Reduction

Each of the algorithms in this chapter attempted to quantize the data by coming up with another point that is reasonably close. This scheme can work well in many cases. Another approach is to simply reduce the dimension of the problem. That is, remove dimensions to save space.

The simplest way to understand this is to think about a two-dimensional picture of a three dimensional object. This removes one dimension of data, and this may or may not add some confusion. A picture of a person taken from the front does a good job of showing many of the features, but it makes it difficult to determine the size of a person's nose.

The goal in dimension reduction is to strip away unnecessary dimensions and create a version of the data in the lower dimensions that does a good job of representing the data. Of course, the phrase "good job" is pretty ambiguous and subject to interpretation. For the purposes of this section, let this phrase indicate that there is some metric $\delta(x, y)$ such that $\delta(x_i, x_j)$ is close to $\delta(p(x_i), p(x_j))$ for all x_i and x_j and where $p(x)$ is the projection that reduced the dimension.

One mathematical operation that accomplishes this job is known as the *singular value decomposition*. It takes an $m \times n$ matrix A and produces two square matrices, U and V and another $m \times n$ diagonal matrix D such that $A = UDV^t$. U is $m \times m$, and V is $n \times n$. Both U and V are orthogonal matrices.

The elements along the diagonal of D are known as the *singular values* of the matrix A. These are usually denoted as $\sigma_1 \geq \sigma_2 \geq \ldots \geq \sigma_k$, where $k = min(m, n)$. The size of these values generally indicate the size of the data in the rows of A.

One way of thinking of the singular value decomposition is to imagine fitting an ellipsoid around the cloud of data. Let each data point be a vector from the origin, and these make up the rows of A. The size of this ellipsoid is defined by the size of this cloud of data. The singular values are the measure of the size of the different axes of the ellipsoid. If you want to reduce the data to three dimensions, then choose the three largest singular values and look at the ellipsoid from that position.

To be more precise, let j be the number of desired dimensions after projection. Set $\sigma_i = 0$ for all $i > j$, and construct D' with these zeroed out singular values. UD' is the result of projecting these values down to j dimensions. V^t is just a rotation matrix, and it can be ignored.

Figure 8.5 shows the result of looking at six clusters of three points in only two dimensions. It is possible to estimate the maximum "error" in the drawing by

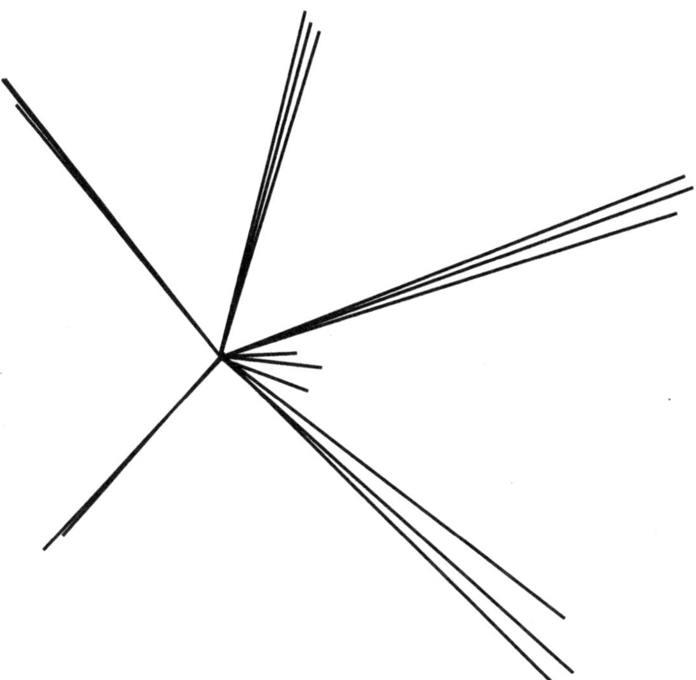

Figure 8.5: The result of projecting a six clusters of three points from their original ten dimensions down to two. There is about 16% error.

looking at the singular values

$$\frac{\sum_{i=1}^{j} \sigma_i}{\sum_{i=1}^{k} \sigma_i}.$$

8.5 Conclusion

This chapter described how to reduce the size of a file by "quantizing" the data set. That is, a set of n values, $d_1, \ldots, d_n$ were chosen as representatives, and the file was compressed by replacing each item with the closest representative. There are a number of different ways for choosing this set of representatives, but they can be summarized as trying to choose either a set that is spread out over the entire space or one that is concentrated where the points might be.

Choosing a set that is spread out over the entire region does a good job of minimizing the maximum error. It ensures that every value in the file is not too far from one representative. This is a good approach for files with data that is fairly evenly distributed over the entire space. It also has good average performance in these cases.

If the data in the files falls into clusters, it can make sense to choose representatives that are also bunched in these clusters. This will ensure that the average distance is small for most of the items. Some outlying data points, however, can be ill represented by this approach, and this may be glaring in image compression.

The most common use for quantization is image compression, because there are many images being shipped over the Internet. One of the most common forms is the Graphics Interchange Format (GIF). This provides a standard way to ship a table of representative colors known as a palette along with a list of pixels that point to the representative.

Chapter 9

Wavelet Transforms

One of the most popular forms of sound or video compressing signals is known as the *wavelet transform*. This is a fairly simple idea that can require a fair degree of mathematical sophistication to implement. It is a popular approach, however, and many tool kits can be found that are designed to make it simple. In fact, there is an entire class of special processing chips known as digital signal processors, or DSPs, that have special multiplying circuits to make it possible to do this form of compression quickly. [1]

Some of the wavelets in this chapter were built, tested, and graphed using Mathematica and its Wavelet Explorer [He96].

The basic conceit is simple: Choose a set of representative patterns and compress a signal by finding a subset of these patterns that add up to the signal. For the sake of simplicity, assume that the signal is a one-dimensional function $f(t)$, where t measures time. This might be a sound file that stores the intensity of the sound every millisecond. There are also two-dimensional approaches that are more appropriate for image compression, and these are described later. Their basic behavior, however, is very similar.

Now, let $g_i(t)$ be another set of functions known as the *basis*. This is a mathematical term that means it covers the entire space of possible functions. That is, any function f can be represented as the weighted sum of some set of these g_i. In other symbols,

$$f(t) = \sum_i \alpha_i g_i(t).$$

[1] Books by Charles Chui [Chu92a, Chu92b, Chu97], Albert Cohen and Robert Ryan [CR95], Ingrid Daubechies [Dau92], Gerald Kaiser [Kai94], A. K. Louis, P. Maaß, and A. Rieder [LMR97], Yves Meyer [Mey92, Mey93], and Gilbert Strang and Truong Nguyen [SN96] are all solid, book-length surveys of the field.

The values of α_i are known as the coefficients. They can be either boolean values, integers, or, in most cases, real numbers. The whole process of compressing a signal boils down to converting a signal f into a set of coefficients $\{\alpha_1, \ldots, \alpha_n\}$. Both the compressing software and the decompressing software share the same set of functions, and the set of coefficients serves as the compressed file.

Once a compression function has found a set of applicable coefficients, the game becomes finding an ideal way to represent the coefficients in the smallest amount of space. The easiest solution is to round off the coefficients, which probably were double precision real numbers, through a process known as quantization. (See Chapter 8.) Small coefficients can usually be reduced to zero with little degradation of the signal. Larger ones can be converted into the nearest one of a set of prearranged points. It is common, for instance, to choose 256 values that do a good job of approximating the potential coefficients and replace the coefficient with an eight-bit number. [2]

The biggest challenge is finding a good set of basis functions for a particular class of signals. No set is perfect for all signals, but some are generally very good. Some excel for certain types like the human voice, which uses a narrow range of frequencies, while others do a better job with sounds like music that encompass a wide range of frequencies. Often, the choice of basis functions is made as much by art as a science, and people choose the one solution that seems to sound best to their ears.

In many cases, the processes of choosing the basis functions and tweaking the quantization routines are tightly integrated. Many of the best audio compression routines, for instance, use different types of quantization for different frequencies. The most important frequencies are the ones to which the human ear is most sensitive. These are quantized with a small quantum to ensure good reproduction. Less important frequencies may simply be rounded to zero in most cases because the ear might not hear them well in any case. The process of creating these quantization tables is as much an art as a science, and the creators often spend long hours conducting blind (or deaf) trials comparing results for fidelity and quality.

Some of the most common basis functions are the sine and the cosine functions. This is partial historical. J. B. J. Fourier was the first to work in this area when he proved that every continuous function can be represented as the sum of sines and cosines. The catch is that he needed to add up an infinite number of sines and cosines to represent each function exactly. Of course, replacing each signal with an infinite number of coefficients is not a particularily efficient compression

[2] [CGV96, KLH98, LRO97, PMN96, SdSG94, WCMP96] offer insight into how to use vector quantization in wavelet and other coding solutions.

technique.

In the many years following, many people have studied the Fourier's idea and come up with a number of different forms that are both practical to compute and very useful in many situations. These are now the foundation for several popular image-compression standards, such as JPEG. The most common solution today is known as the *Discrete Fourier Transform*, which computes the set of coefficients when the function $f(t)$ is described for a fixed number of discrete points.

The JPEG standard is described in Chapter 10. Some audio standards described in Chapter 12 also use it.

In recent years, mathematicians have begun branching out and experimenting with more complicated sets of basis functions. The term *wavelet* is used to encompass all of these possible functions, which may take stranger and more arcane shapes. Much of the research in recent years focussed on narrowing the size of each wave so that it only applies to a certain part of the signal. It is common, for instance, to multiply the basis functions by a packet that is zero outside of some narrow range, say between 10 and 11. This process finds a different set of coefficients for this narrow range and is repeated for all of the important ranges that need to be compressed. This tight focus is why the term "wavelet" became popular, because it captures how the compression accomplished by tiny waves, not long infinite ones.

The rest of this chapter will examine the process of finding the sets of basis functions and calculating the coefficients that for particular signals. The first part will concentrate on one-dimensional solutions, which are useful in audio signals. This is also an easier domain for introducing the concepts. The second part will examine the two-dimensional solutions.

[BZM98a, BZM98b, CGO94] give a few examples of how to use wavelets in medical imaging without losing detail.

9.1 Basic Fourier Mathematics

Fourier mathematics is based upon continuous functions and symbolic mathematics. It is not generally suited to compressing actual data files, which are only discrete collections of points, not symbolic representations of data.[3] Still, this brief introduction helps explain some of the intellectual foundations, and so it might help provide some context for why people came up with the solutions that they use. They were not generated out of thin air but came about as the field evolved.

Let $f(t)$ be a function defined over the range $0 \leq t \leq 2v$. The goal is to find

[3]MIDI files contain music represented as symbolic sums of waveforms. They make a good compression solution, but there is no easy way to take regular recordings and convert them into MIDI files.

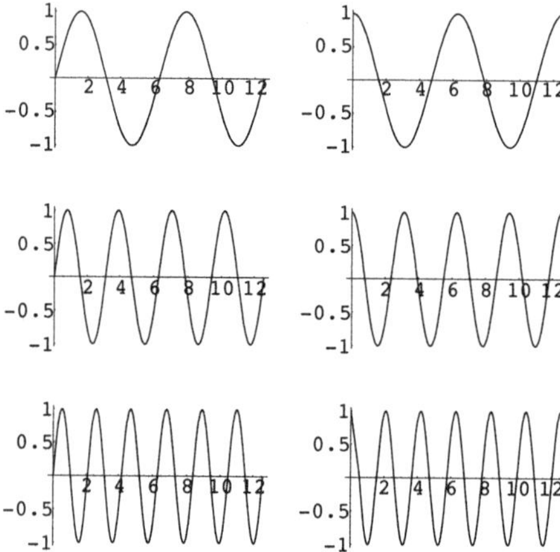

Figure 9.1: Here are the first three basis functions for sine and cosine, where $j = 1, 2,$ and 3. In this case, $v = \pi$.

a set of coefficients c_j and d_j such that

$$f(t) = \frac{c_0}{2} + \sum_{j=1}^{\infty} c_j sin(\frac{j\pi t}{v}) + d_j cos(\frac{j\pi t}{v}).$$

The basis functions are $sin(\frac{j\pi t}{v})$ and $cos(\frac{j\pi t}{v})$. Figure 9.1 shows the first three functions. The coeffients c_j and d_j can be calculated with these integral equations:

$$c_j = \frac{1}{v}\int_0^{2v} f(t)cos(\frac{j\pi t}{v})dt \quad d_j = \frac{1}{v}\int_0^{2v} f(t)sin(\frac{j\pi t}{v})dt.$$

These equations look fairly complicated, but there is a simple explanation. The coefficients are determined by the amount of overlap between the basis functions and f. The integrals calculate the average overlap for the entire span between 0 and $2v$.

Figure 9.2 shows an example where a simple function, $f(t) = sin(2\pi t) + .2sin(4\pi t) + .3sin(8\pi t)$, is decomposed into coefficients by multiplying it by $sin(2\pi t)$, $sin(4\pi t)$, $sin(6\pi t)$, and $sin(8\pi t)$. The result of the four integrals are $1, 0, .2$ and $.3$, respectively. Of course this example is fairly contrived so the numbers work out exactly. In other cases, there will be a wide range of values.

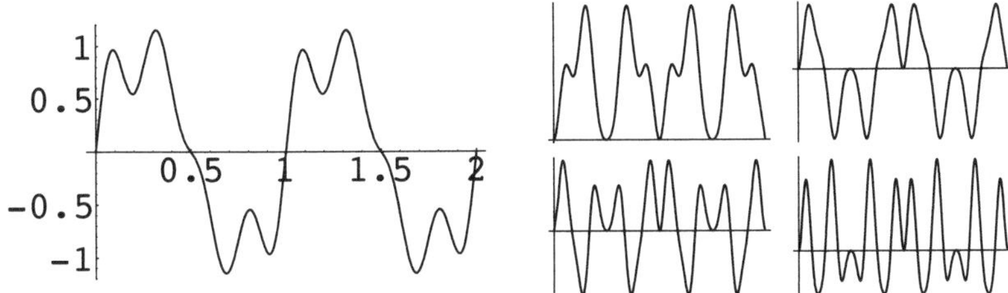

Figure 9.2: Here is a graph of the function $f(t) = sin(2\pi t) + .2sin(4\pi t) + .3sin(8\pi t)$ on the left and four graphs of the integrals of f with $sin(2\pi t), sin(4\pi t), sin(6\pi t)$, and $sin(8\pi t)$ on the right. If the four graphs on the right are integrated, the results returns the Fourier coefficients.

This example also illustrates an interesting property of the functions Fourier choose as a basis. They are *orthogonal*. This has a formal definition: $g_i(t)$ and $g_j(t)$ are orthogonal if

$$\int_0^{2v} g_i(t)g_j(t) = 0, i \neq j.$$

The functions are also *normalized*, because

$$\int_0^{2v} g_i(t) = 1.$$

This is true for Fourier's basis functions, $sin(\frac{j\pi t}{v})$ and $cos(\frac{j\pi t}{v})$.

Normalized, orthogonal collections of basis functions are easy to use because they provide a unique set of coefficients. They are also easier to calculate and make it possible to take short cuts.

There are several interesting features of these equations. The coefficients, c_j and d_j, act as measures of frequences. If c_j is large, then there is a pattern in f with frequency $\frac{j\pi}{v}$. For this reason, some people refer the collection of coefficients, $\{c_0, \ldots, c_n\}$ and $\{d_1, \ldots, d_n\}$ as the *frequency space*. The values summarize the frequencies found or not found in the signal. It should be clear why many radio engineers who want to pluck signals of particular frequencies out of the air spend plenty of time using Fourier transforms. Cell phones are just computers that do Fourier transforms all of the time.

Fourier originally discovered that this collection of functions could be set to equal all properly behaved functions that were continuous with at least two continuous derivatives. Exact equality, of course, depended upon using an infinite number of terms. The rest of this section will concentrate on finding ways to get by with the fewest number of coefficients in order to increase the amount of compression.

9.2 Discrete Cosine Transform

One of the most common forms of the Fourier approach today is known as the *discrete cosine transform*. It is called *discrete* because the function f that is approximated is only defined at a limited number of points, usually a power of 2 like 8, 64, or 128. It is called a *cosine* transform because the basis functions are only cosines. This adds a certain amount of simplicity.

The DCT is a widely studied extension of the Fast Fourier Transform. [ANR74, CL91, CSF77, CBSC94, Hou87, MS93, NP78, WP91] offer deeper insight than this book.

The discrete cosine transform or DCT is quite popular today because it is the foundation for many of the different image-compression functions that are commonly used. The JPEG standard uses the DCT for heavy compression, and many of the other video formats also use it in some form. While many people continue to experiment with other collections of basis functions and wavelets, the DCT is one of the most common today.

While the regular Fourier transform is calculated with integrals, the discrete version replaces the process of integration with summation. These are equivalent operations, and the approach is the same in both cases. In general, it is easier to use matrix and vector notation to describe the DCT because it is both familiar and compact.

Chapter 10 describes the use of the DCT in the JPEG standard. See also [PM93, Wal91].

In the interest of completeness, here is a quick summary of matrix notation. A n-dimensional *vector* is a group of n numbers that are usually stacked vertically like this:

$$\begin{matrix} v_0 \\ v_1 \\ v_2 \\ \vdots \\ v_{n-1} \end{matrix}$$

This is usually called v, but some books use $\bar{v}$ to distinguish vectors from scalars.

A matrix is an $m \times n$ collection of numbers arranged in a block. They usually look like this:

$$
\begin{array}{lllll}
M_{0,0} & M_{0,1} & M_{0,2} & \cdots & M_{0,n-1} \\
M_{1,0} & M_{1,1} & M_{1,2} & \cdots & M_{1,n-1} \\
M_{2,0} & M_{2,1} & M_{2,2} & \cdots & M_{2,n-1} \\
\vdots & & & & \vdots \\
M_{m-1,0} & M_{m-1,1} & M_{m-1,2} & \cdots & M_{m-1,n-1}
\end{array}
$$

A vector can be multiplied by a scalar, Mv, and this will produce a new vector. The result is pretty straightforward, if a bit repetitive. In this case, the matrix M is sliced into rows, which are each combined with v in turn using a process known as the *dot product*. The dot product is calculated by adding up the result of multiplying the individual elements together. This may be better explained by a formula

This introduction to matrix and vector manipulation leaves out a number of important details. A deeper treatment can be found in many vector algebra books.

$$w_j = \sum_{i=0}^{n-1} M_{j,i} v_i.$$

The values of w_j form a new vector.

In a DCT, the data to be compressed are expressed as a vector: v. If the signal is an audio signal, then it probably measures the intensity at n different time intervals. The process of converting the data elements into "frequency elements" is achieved by multiplying by a matrix M. The elements in this matrix look like chunky versions of cosine functions.

The dot product is equivalent to the integral illustrated in Figure 9.2. It is equivalent to determining how much two vectors "overlap".

In the DCT commonly used, the term at position (i, j) of matrix M is

$$c_i cos[\frac{(2j+1)i\pi}{2N}],$$

where N is the is the number of rows and columns in the matrix. $c_0 = \frac{1}{\sqrt{2}}$ and $c_i = 1$ for all other i.

Here's what the matrix for $N = 8$, call it M_8, looks like:

$$\begin{matrix}
\frac{1}{\sqrt{2}} & \frac{1}{\sqrt{2}} & \frac{1}{\sqrt{2}} & \frac{1}{\sqrt{2}} & \frac{1}{\sqrt{2}} & \frac{1}{\sqrt{2}} & \frac{1}{\sqrt{2}} & \frac{1}{\sqrt{2}} \\
\cos\frac{\pi}{16} & \cos\frac{3\pi}{16} & \cos\frac{5\pi}{16} & \cos\frac{7\pi}{16} & \cos\frac{9\pi}{16} & \cos\frac{11\pi}{16} & \cos\frac{13\pi}{16} & \cos\frac{15\pi}{16} \\
\cos\frac{2\pi}{16} & \cos\frac{6\pi}{16} & \cos\frac{10\pi}{16} & \cos\frac{14\pi}{16} & \cos\frac{18\pi}{16} & \cos\frac{22\pi}{16} & \cos\frac{26\pi}{16} & \cos\frac{30\pi}{16} \\
\cos\frac{3\pi}{16} & \cos\frac{9\pi}{16} & \cos\frac{15\pi}{16} & \cos\frac{21\pi}{16} & \cos\frac{27\pi}{16} & \cos\frac{1\pi}{16} & \cos\frac{7\pi}{16} & \cos\frac{13\pi}{16} \\
\cos\frac{4\pi}{16} & \cos\frac{12\pi}{16} & \cos\frac{20\pi}{16} & \cos\frac{28\pi}{16} & \cos\frac{4\pi}{16} & \cos\frac{12\pi}{16} & \cos\frac{20\pi}{16} & \cos\frac{28\pi}{16} \\
\cos\frac{5\pi}{16} & \cos\frac{15\pi}{16} & \cos\frac{25\pi}{16} & \cos\frac{3\pi}{16} & \cos\frac{13\pi}{16} & \cos\frac{23\pi}{16} & \cos\frac{1\pi}{16} & \cos\frac{11\pi}{16} \\
\cos\frac{6\pi}{16} & \cos\frac{18\pi}{16} & \cos\frac{30\pi}{16} & \cos\frac{10\pi}{16} & \cos\frac{22\pi}{16} & \cos\frac{2\pi}{16} & \cos\frac{14\pi}{16} & \cos\frac{26\pi}{16} \\
\cos\frac{7\pi}{16} & \cos\frac{21\pi}{16} & \cos\frac{3\pi}{16} & \cos\frac{17\pi}{16} & \cos\frac{31\pi}{16} & \cos\frac{13\pi}{16} & \cos\frac{27\pi}{16} & \cos\frac{9\pi}{16}
\end{matrix}$$

The JPEG and MPEG algorithms use the 8×8 DCT extensively. See Chapters 10 and 11.

The rows of this matrix are orthogonal and normalized. An interesting feature of these matrices is that they are recursively defined. M_4, for instance, is a subset of M_8. These systems were designed to make it simple to calculate the results.

A quick glance at the matrix shows plenty of patterns that can be extracted. Fourier transforms spawned the Fast Fourier Transform, which has a cousin, the Fast Discrete Cosine Tranform. Each of these approaches exploits the patterns in matrix to reduce the complexity to $O(n \log n)$, where n is the number of points being fitted.

9.3 Two-Dimensional Approaches

The one-dimensional algorithms described above are useful for compressing audio signals. They are also easy to understand. Both Fourier and discrete cosine transforms have been extended into two dimensions, and these algorithms are used quite frequently to compress two-dimensional images.

The two-dimensional versions are simple extensions. They are often cast as sums of exponential functions. This representation is cleaner and relies upon the fact that

$$e^{\pi i x} = cos(\pi x) + isin(\pi x).$$

The discrete version uses N^2 different functions to calculate the frequency space of an N^2 image where the intensity at pixel (p, r) is represented as $a_{p,r}$. The equation for calculating the frequency terms for this discrete form is

$$b_{s,t} = \frac{1}{\sqrt{n}} \sum_{p=1}^{n} \sum_{r=1}^{n} a_{p,r}(e^{2\pi i(p-1)(s-1)/n} + e^{2\pi i(r-1)(t-1)/n}).$$

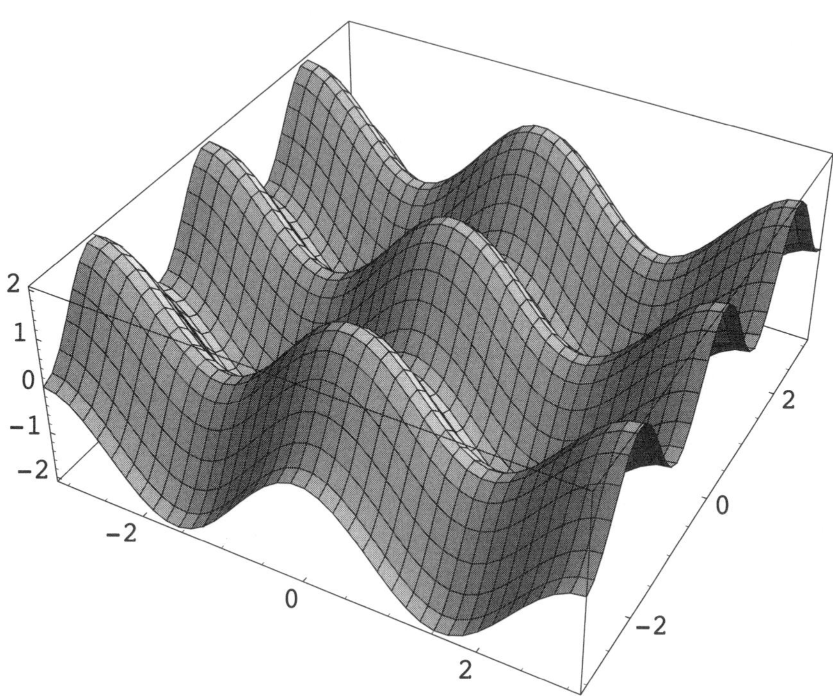

Figure 9.3: One of the many two-dimensional wave functions used to create the two-dimensional Fourier transform.

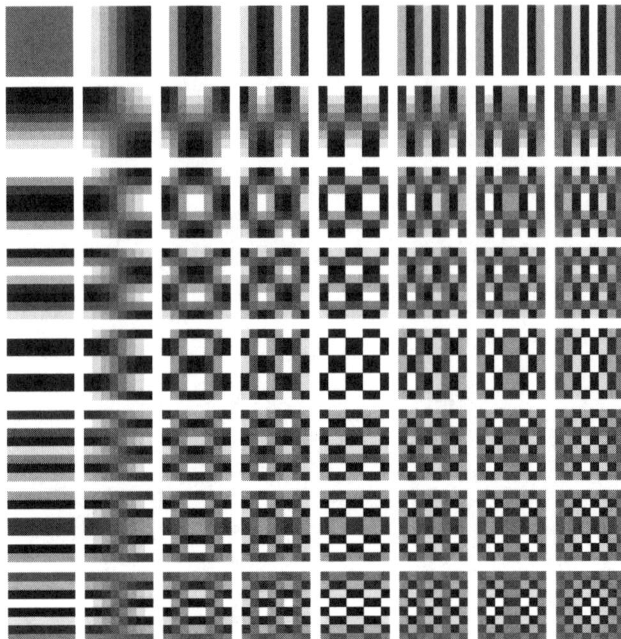

Figure 9.4: The 64 different two-dimensional basis functions used in the two-dimensional discrete cosine transform of 8×8 grid of pixels. The intensity at each particular point indicates the size of the function.

Figure 9.3 shows the real part of $e^{2ix} + e^{3iy}$. The discrete Fourier function must be calculated with complex numbers that have both real and imaginary parts. The discrete cosine function is a simplification that uses only real values. This is generally acceptable for practical purposes because most real-world data does not come with an imaginary component.

Figure 9.4 shows all of the 64 different basis functions used in the two-dimensional discrete cosine transform of an 8×8 grid of pixels. The functions are created by oscillating the function at different frequencies along the two different axes. This shows all possible combinations.

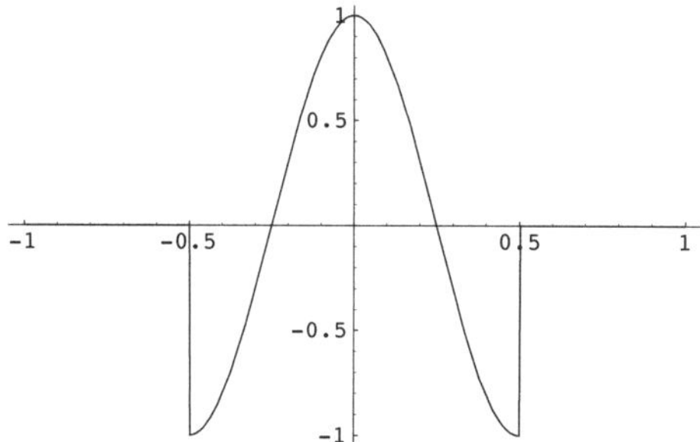

Figure 9.5: The function $cos(2\pi x)$ is cut to zero outside of the range $(-.5, .5)$.

9.4 Other Wavelet Functions

Of course, sines and cosines aren't the only basis functions used with the wavelet transform. They were just the first and the most common. Lately, the area has attracted a great deal of study by mathematicians who found a wide variety of other functions that can do a good job.

The functions established by Fourier set a high standard for the new mathematicians to surpass. The sine and cosine functions are fairly easy to describe, orthogonal, normalized, and well understood mathematically. Still, they are not perfect. The biggest problem is that they're highly regular. They repeat every 2π units — something that's okay for repeated functions but problematic for ones that change.

The best solution for real-world data is to create localized versions of the sines and cosines, that is, functions that have strong peaks in narrow regions. These functions can be scaled with coefficients and moved to the right location to hit the peaks. This approach can be more efficient because the localized waves, hence *wavelets*, are responsible only for mapping out a small region of the function.

Figure 9.5 shows a truncated cosine function designed to be more localizable. This version is created by multiplying a standard cosine function by a *window function* that chops the size of the function to zero outside of the range $\{-\frac{1}{2}, \frac{1}{2}\}$. Figure 9.6 shows the same function, $cos(2\pi x)$, being truncated by a window function defined to taper gently to zero outside the range $\{-\frac{1}{2}, \frac{1}{2}\}$. The slope of the truncation can be defined to make the transforms converge nicely. The actual description of the slopes is beyond the scope of this book.

Another popular approach used in wavelet analysis is *dilation*. In this case, the function might be said to be dilated to be one unit long. This step is combined with

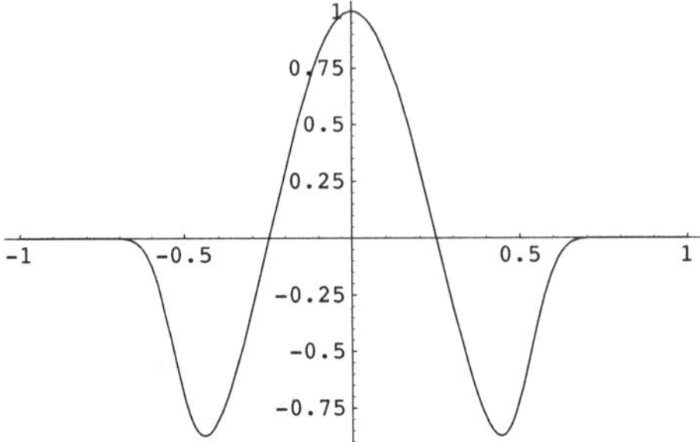

Figure 9.6: The function $cos(2\pi)$ function gently truncated to zero outside of the range $(-.5, .5)$.

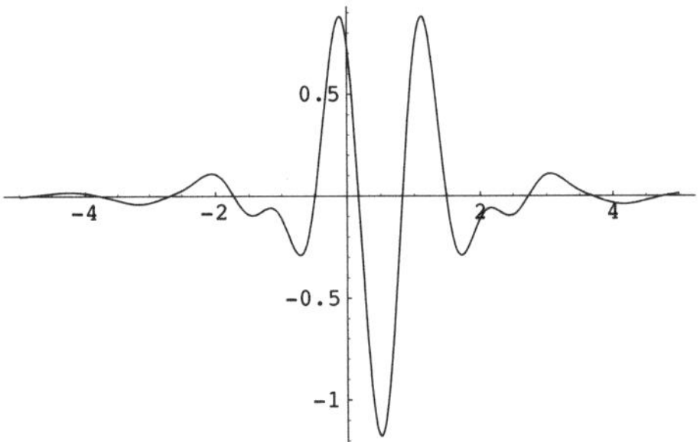

Figure 9.7: A Meyer $\Psi(x)$ function used as a wavelet.

raw translation to assemble a set of wavelets that will replicate a function when translated to the right location, dilated to the right width, scaled by coefficients to be the right height, and then added together. The compressed version of the function can be represented by the coefficients, translation amounts, and dilation factors.

[Tha96, ZY96] combines fractal compression (Chapter 13) with the discrete cosine transform.

Most of the mathematical foundations of wavelet analysis are beyond the scope of this book. There is a great deal of clean, elegant mathematics that explains how and why to design wavelets to be orthogonal, normalized, and efficient functions that are easy to use in computations. This insight can be powerful for mathematicians, but much of it is not practical for the average user of compression functions.

Figure 9.7 and 9.8 show some sample wavelets, the first developed by Yves Meyers and the second constructed by Ingrid Daubechies.

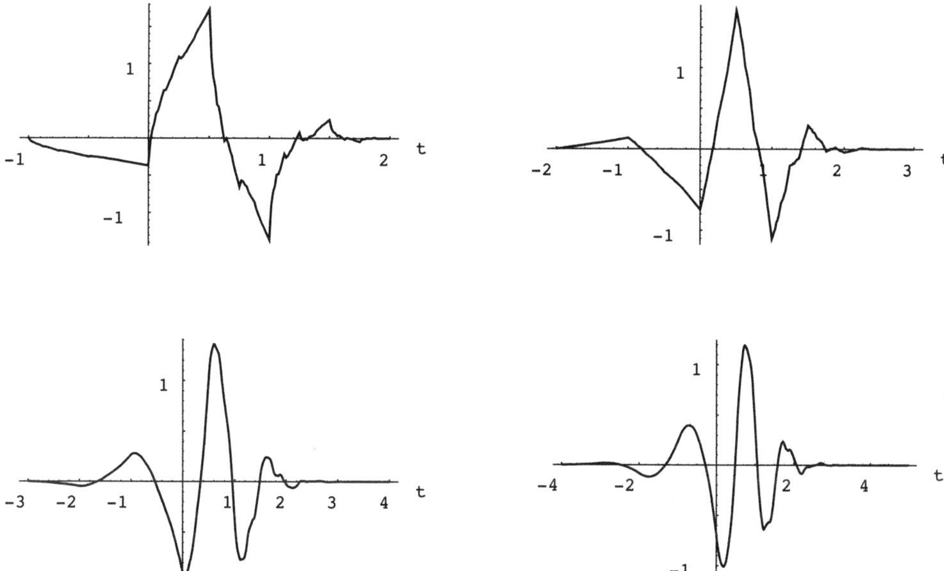

Figure 9.8: Four Daubechies $\Psi(x)$ functions, which are often used as wavelets.

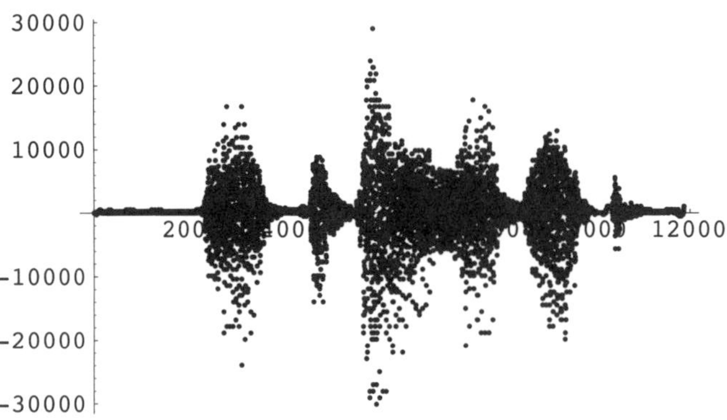

Figure 9.9: A graph of a sound file where the x-axis measures time and the y-axis measure intensity.

Multiple Resolution Analysis

One of the more effective ideas is to use multiple resolutions to scale and transform wavelets so that the function is modeled as effectively as possible. This might be accomplished by first using calculating the coefficients necessary to model a function in the range $\{0, 1\}$. Then, when this doesn't fit, split the range in half and scale the wavelets to model the function in $\{0, \frac{1}{2}\}$ first and then $\{\frac{1}{2}, 1\}$ afterwards.

This process of splitting a region in half and repeating the process at a finer scale can be repeated as many times as is necessary. Some smooth, relatively regular functions won't need to be analyzed in so many steps. Other irregular ones might need some regions split up more than others. There is no reason why one region needs to be cut up as much as others.

Figure 9.9 shows a sound file created from a microphone. It should be obvious that it can't be easily replicated by simple cosine functions, because the peaks are irregularly spaced and of different sizes. It was fit with a set of windowed cosine packets like the ones shown in Figure 9.5. When the match wasn't too good, the region was split in half and matched separately. Figure 9.10 shows the sizes of the packets used to model the function. In the first round, the sound file was split into 11 different regions of the maximal size. The first two regions were mainly silence so they were easily modeled, and there was no need to split them further. This is represented by filling in the rectangles in the second row. Other segments have

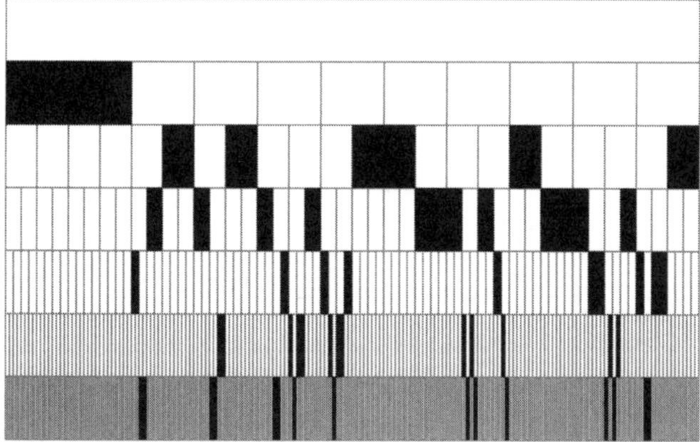

Figure 9.10: This shows which size wavelets were used in a discrete cosine transform of the sound file shown in Figure 9.9. The dark section shows the smallest wavelets used to represent that particular segment. For instance, the first part of the signal is mainly silence, so a few large wavelets model the signal well. The most complicated parts of the signal are the peaks where the intensity changes frequently. In these cases, the wavelet transform uses the greatest amount of recursive decomposition, a fact that is indicated by the blocks colored in at the bottom of the graph.

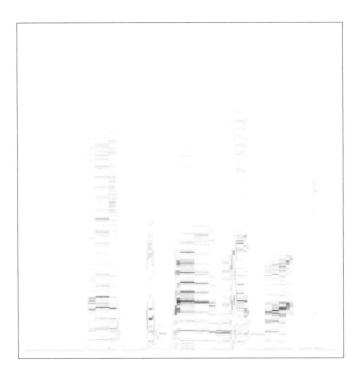

Figure 9.11: The frequencies produced by the discrete cosine transform. The x-axis shows the time, and the y-axis shows the frequency. The strength of the coefficient is shown by the intensity of the plot.

rectangles filled in lower on the scale. This indicates that the wavelet transform modelling system needed to break up the regions into smaller regions to do a better job of representing the underlying sound file.

This file could be compressed by zeroing out the small coefficients and quantizing the other ones to be represented with a smaller amount of space.

A 2D Example of Multiple Resolution Analysis

The multiple resolution decomposition of an image is a particularily effective way to compress it, because many parts of an image are often free of detail. That is,

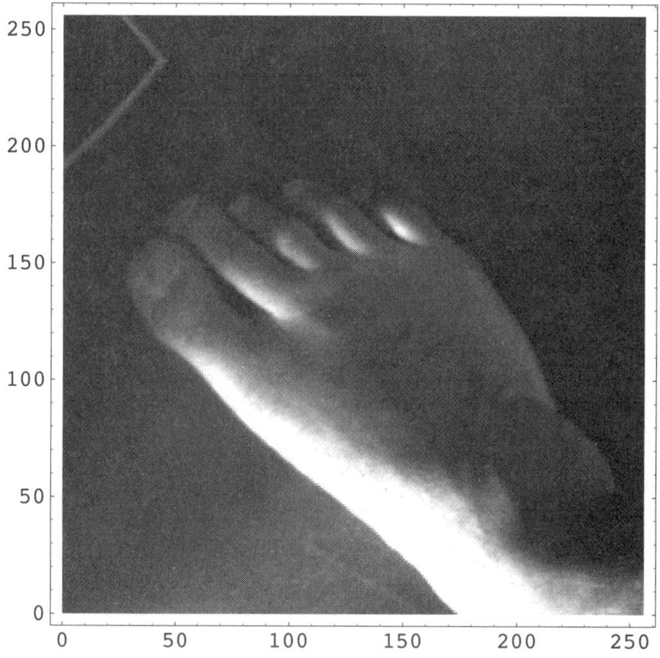

Figure 9.12: A picture of the author's foot.

many parts of an image contain large blocks of pixels with close to the same value. It makes sense to try to represent all of them with one number.

This file could be compressed by zeroing out the small coefficients and quantizing the other ones to be represented with a smaller amount of space. Figure 9.12 shows a picture of the author's foot that is about to be compressed using Daubechies's least asymetric filter and the wavelet functions built into Mathematica's Wavelet Explorer.

Figure 9.13 illustrates the results of computing this wavelet transform. This was done at three different levels of granularity: $32 \times 32, 64 \times 64,$ and 128×128. The first row shows a 32×32 grid of the *residual values*, that is, the average value of the 8×8 groups of pixels. The second row shows the coefficients for three 32×32. These are the coefficients for the Ψ^{I},Ψ^{II}, and Ψ^{III} functions. The third row shows the coefficients from fitting the same Ψ^{I},Ψ^{II}, and Ψ^{III} at lower granularity of blocks of 4×4 pixels at a time. The final row shows the same

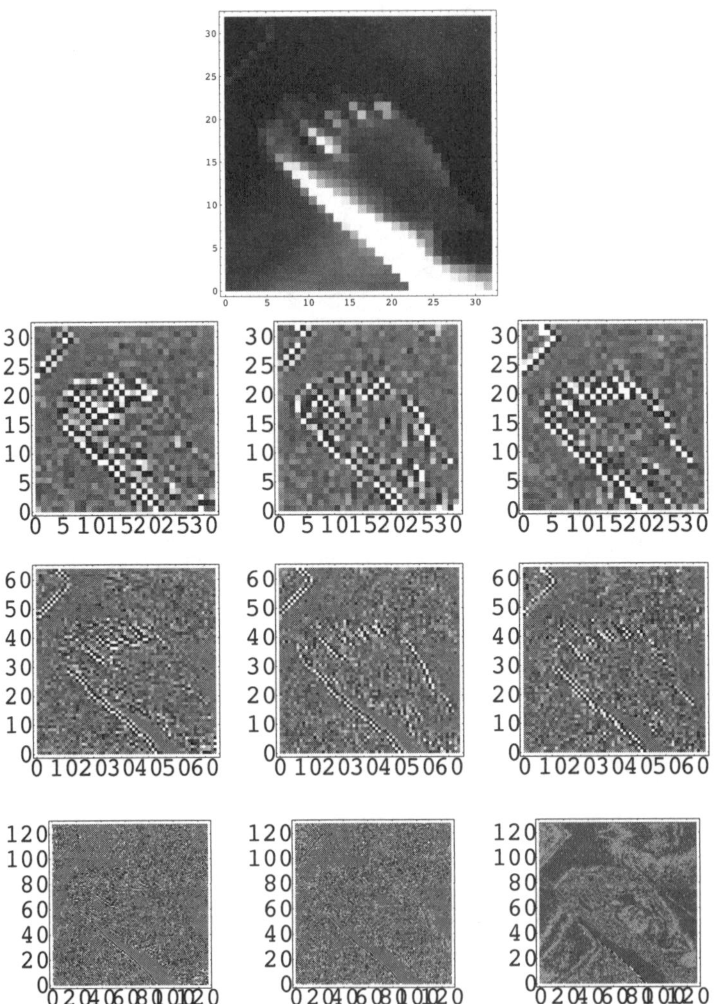

Figure 9.13: The coefficients of the wavelet transform of Figure 9.12.

functions fit to 2 × 2 blocks of pixels.

The definitions of Ψ^I, Ψ^{II}, and Ψ^{III} are beyond the scope of this book.

Figure 9.14 shows the results of doing an inverse wavelet transform on the coefficients captured in Figure 9.13. The first row, left drawing shows the results of doing an inverse transform on just the 32 × 32 data. The right drawing shows the extra detail hidden in the 64 × 64 data. The bottom row shows them added together. In essence, the 32 × 32 coefficients model the bigger details, while the 64 × 64 grids model the finer details. The 128 × 128 data are not added into this image, but they already do a good job of reconstituting the figure.

The compression is completed by quantizing the most significant coefficients and zeroing out the small or insignificant ones.

A 2D example of the Discrete Cosine Transform

The discrete cosine transform is one of the most common forms of compressing images, thanks to the popularity of the JPEG (Chapter 10) and MPEG (Chapter 11) standards. These systems don't begin to utilize all of the power or capability of the algorithm, because they are limited to avoid overtaxing embedded processors and earlier computers. This example takes the same 256 × 256 pixel image from Figure 9.12 and compresses it in 64 × 64 pixel blocks. The JPEG and MPEG standards use 8 × 8 pixel blocks, which are easier to compute. The smaller block size reduces the complexity of the fast DCT operation and makes it possible for more information to be kept in the cache of a microprocessor. It also reduces the number of circuits that are used in a strict hardware implementation.

One advantage of using larger blocks is that the effects of the compression algorithm are easier to see. Figure 9.15 and 9.16 both show the result of decompressing the data after they have been compressed with the discrete cosine transform. Both make it clear that the edge effects around the boundary of the 64 × 64 pixel block can be significant and often create the greatest visual disturbances. The brain is very good at picking up patterns, and the anolomies that lie in a grid are much more intrusive than those that are randomly distributed.

The larger blocks also make it easy to see some of the wavy errors that can be introduced in heavily compressed images. The discrete cosine transforms are, of course, wavy, and this can often show through if only a few coefficients are kept.

The images in Figures 9.15 and 9.16 were created by doing two different discrete cosine transforms on 64 × 64 pixel blocks of the image. The results shown in Figure 9.15 come from a transform created with out any boundary taper that smooths the transistions between blocks. Naturally, the grid of discontinuities is much more obvious in these versions.

The results in Figure 9.16 were produced by a discrete cosine transform that

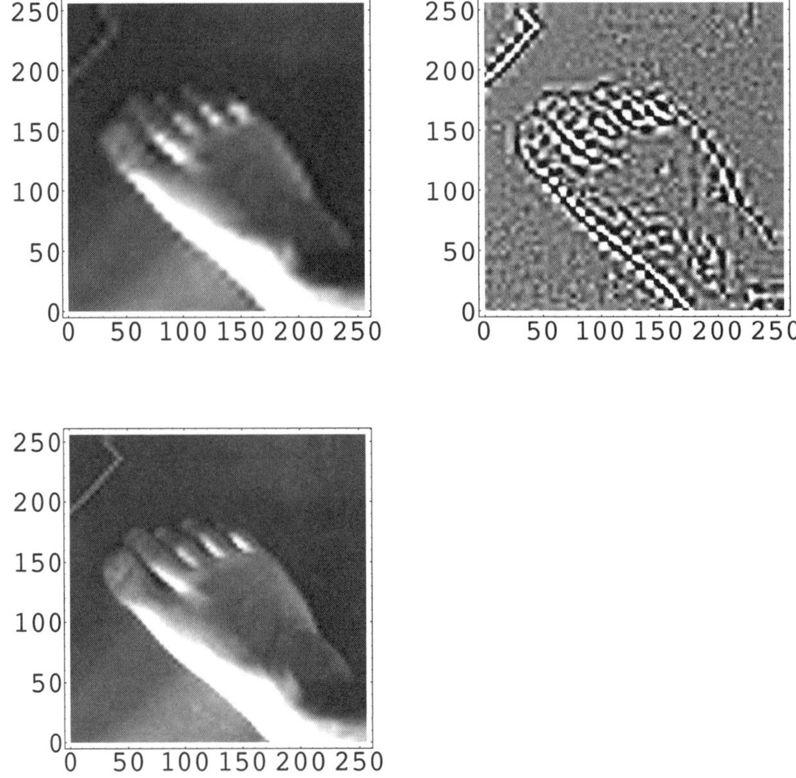

Figure 9.14: The results of doing an inverse wavelet transform. The upper left-hand corner shows the results of using just the 32×32 data. The upper right shows the results from using the 64×64 grid. The bottom shows the sum of both.

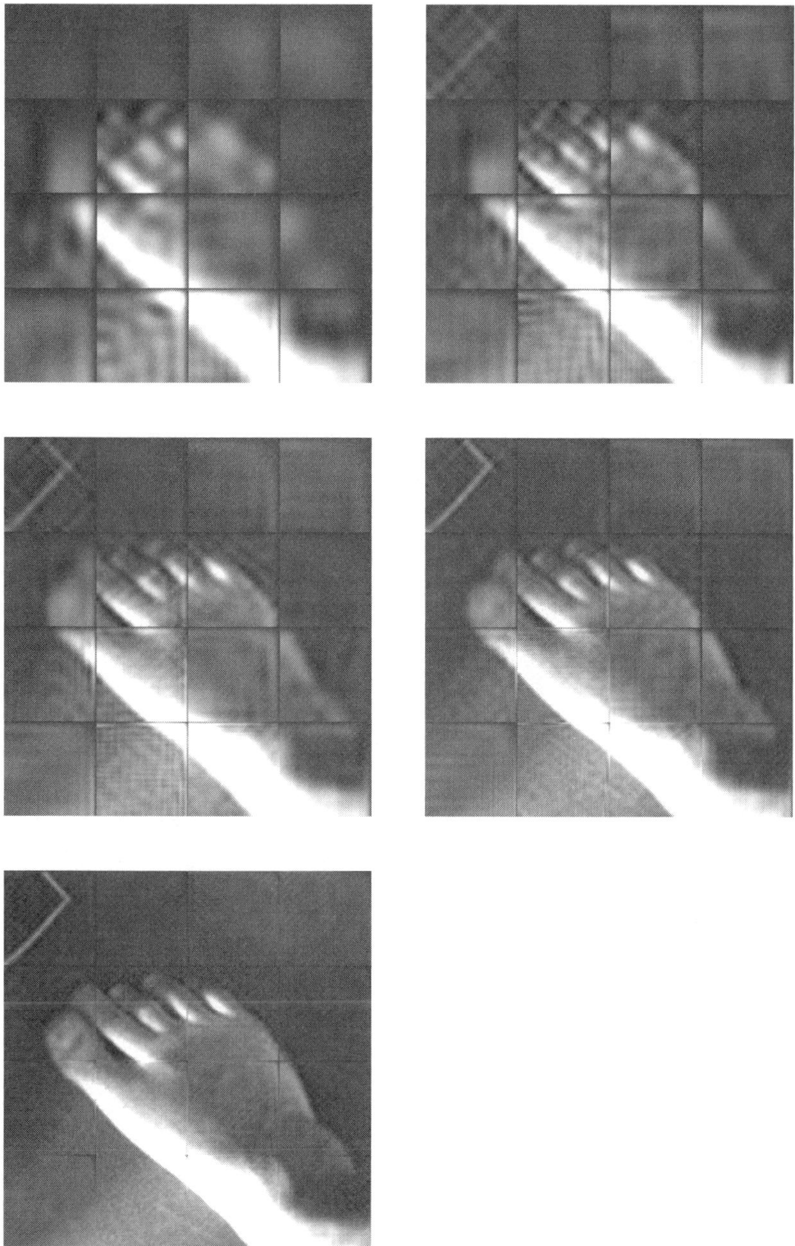

Figure 9.15: The results of doing an inverse cosine transform after keeping 200, 500, 1000, 2000, and 5000 of the largest coefficients. Figure 9.16 shows the same results with a polynomial taper added to reduce the blocking effects.

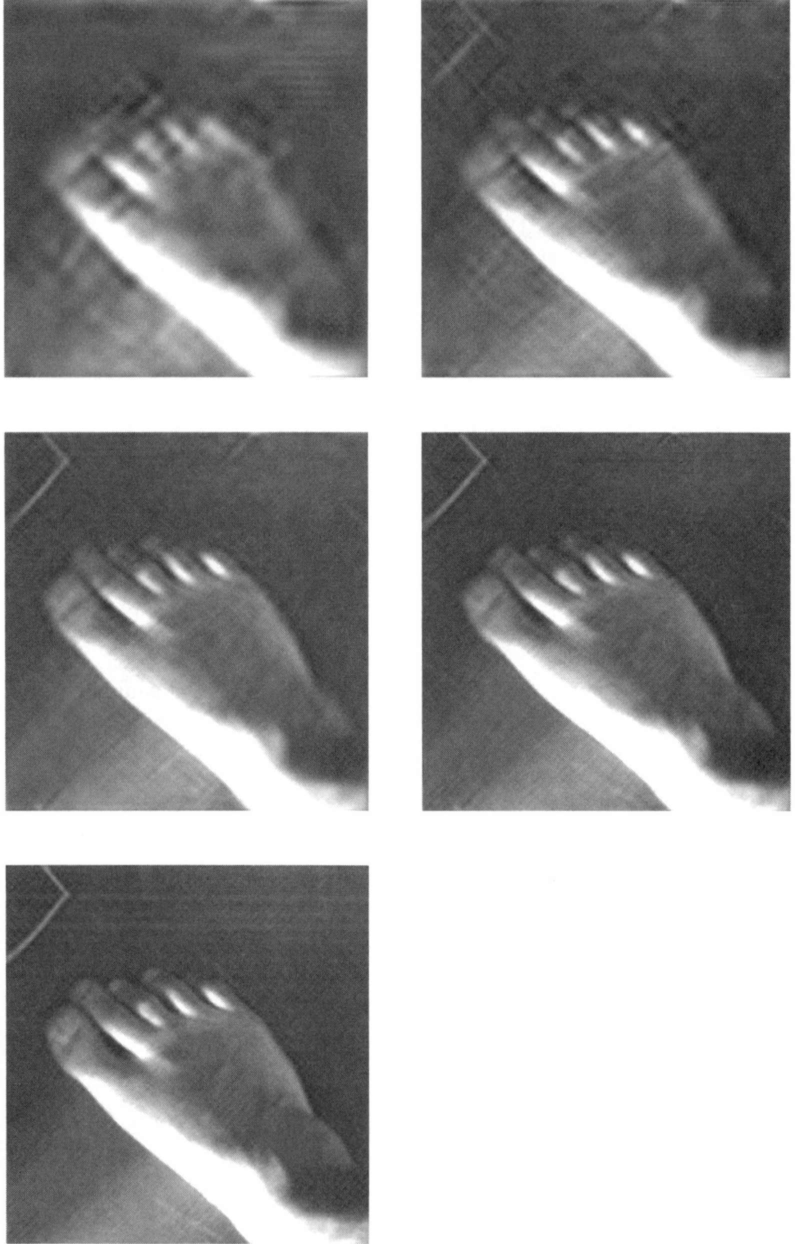

Figure 9.16: The results of doing an inverse cosine transform after keeping 200, 500,1000, 2000, and 5000 of the largest coefficients. In this figure, a polynomial taper was used to reduce blocking effects.

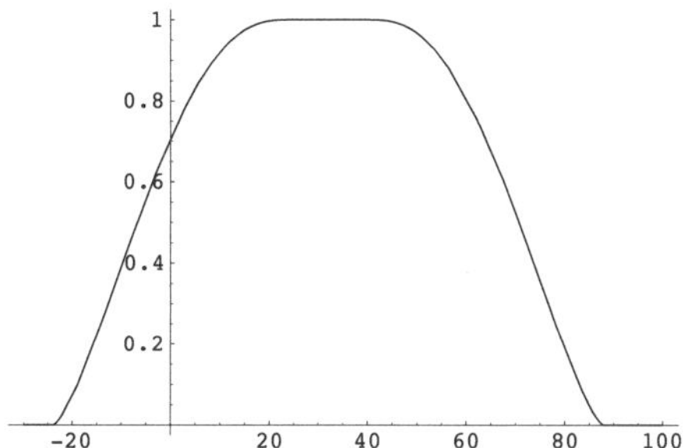

Figure 9.17: A cross section of the taper used to smooth the edges between blocks in Figure 9.16.

was tapered around the edges of the block. This polynomial taper is shown in Figure 9.17. The blurring between blocks removes much of the artifacts created by splitting the image up into blocks.

Both Figures 9.15 and 9.16 were created to show the effects of throwing away coefficients. The images were produced by keeping only the largest n coefficients from the picture where $n = 200, 500, 1000, 2000$, and 5000, when marching from left to right, top to bottom. The worst images are in the top left-hand corner, and they get progressively better as more coefficients are available to add detail.

None of the coefficients in this example were quantized. The side effects in the image are strictly produced by representing the image as the sum of a number of cosine functions.

A quick glance at the images shows that the discrete cosine transform is not as good at reproducing the images when the bit rate is low. The wavelet functions do better because there is some overlap between the adjacent areas. The JPEG committee was aware of these limitations when it chose the 8×8 block DCT for picture compression and made the choice anyway because the DCT was easier to implement in hardware. The patterns in the DCT make it possible to create a fast version of the matrix multiply that takes substantially less time. Also, the DCT is well understood and thus easier for an industry to embrace.

9.5 Conclusion

Wavelet analysis is a popular method of compression that is becoming more sophisticated because it is an active field of research. The goal is to take a set of basic functions and compare them against the data. Each function might be said to predict the data with a particular amount of success, and this success can be measured with a number known as the coefficient. If the basic functions are chosen carefully, then the data can be completely predicted or summarized as the sum of the functions scaled by the coefficients.

Compressing the data with this system is as easy as throwing away many of the coefficients; the more coefficients that are thrown away, the better the compression. The cost of this insouciance, however, is greater error. The goal is to find a good set of functions that models the data with a small number of nonzero coefficients.

Chapter 8 gives a good survey of some of the ways of "throwing away" coefficients. In many cases, the best solution is to quantize the coefficients into one of a few preset values. This takes up little space but comes close enough to the actual value.

[Wat93, MS93, RL95, Saf94] describe better ways of choosing the right quantization levels in order to match human perception.

Some of the best audio systems now try to do adaptive quantization. They look for the most important frequencies and use a sophisticated model of the human ear to allocate the most bits to those frequencies. The less important ones are quantized with less precision or reduced to zero altogether.

The field of wavelet analysis is growing as more mathematicians examine how to create different wavelets for different sets of data. The basic functions like the cosine function continue to be popular with many compression standards because they're so well understood. There's more than a century of both formal analysis and engineering adaption available as guidance. Many of the new wavelets are promising but still unfinished.

Chapter 10

JPEG

The JPEG standard for compressing still photographic images was developed by a committee known as the "Joint Photographic Experts Group", who named the standard after themselves. The solution is one of the most extensive and complicated standards in common use and provides a good example of how all of the techniques described in this book can be combined to produce fairly dramatic compression ratios. It is not unusual to see photographs reduced by ratios of 10:1, and some photographs can be reduced by close to 50:1, albeit with plenty of degradation.

[PM93, Wal91, RH96, Sol97] are other good places to begin learning about the JPEG standard.

The core of the JPEG standard is the discrete cosine transform described in detail in Chapter 9. The transform is applied repeatedly to 8×8 blocks of pixels. In many cases, only the most significant coefficients are kept from this process. This is responsible for much of the compression, but it is also the cause of degradation in the image. JPEG is usually a lossy compression standard that does not reproduce an image exactly. The standard allows the user to control the number of coefficients that are dropped and thus control the tradeoff between saving space and reproducing the images. More compression can be achieved by dropping more coefficients, but this reduces the ability of the discrete cosine transform to represent the data in the 8×8 block of pixels. In some images, a few coefficients may be enough to do the job, but in others, it can lead to a great loss of detail. At high compression ratios, the 8×8 blocks of pixels often blur into solid blocks, and the image can take on a blocky or highly pixelated appearance. The 64 pixels come close to blending together into one big pixel.

The algorithm includes two extra steps that follow the discrete cosine transform and these extra steps can save even more space in the process. The coefficients are not reproduced exactly when they are saved to disk in the JPEG format but are converted into predefined quanta. These values are further compressed by using either Huffman coding or arithmetic coding to pack the coefficients together. The overall standard includes many different variations that use or don't use each of these features. There are lossless versions that use settings that don't discard any information. There are also ones that use different variations of Huffman or arithmetic coding. This chapter will describe the most common variations and also explain how additions and changes can change the system.

See Chapter 8 for information on quantization, Chapter 2 for information Huffman coding, and Chapter 4 for information on arithmetic coding.

10.1 JPEG Overview

There are many different facets to the JPEG compression standard, and it helps to look at each of these facets from a distance before examining the standard in detail. The basic algorithm will just repeatedly push 8×8 pixel blocks through a discrete cosine transform, but there are several different ways that the results can be packaged. The JPEG committee worked toward building a flexible standard that could solve many people's problems at the same time.

The most obvious difference between different applications of the standard is the level of quality. The JPEG group set out to create a flexible standard that would allow the user to trade off quality of reproduction for more compression. The images reproduced by the system are supposed to degrade in a fairly gradual and manageable way. At the high end of the continuum, the standard should reproduce the images exactly. Just below this level, the images should be visually indistinguishable, although they may not be reproduced exactly. The JPEG group set a goal of about two bits per pixel for this level of quality. The quality should degrade slowly as the bits used for each pixel drop. At .25 bits per pixel, the JPEG standard group hopes that the images are still classified as "moderate to good quality".

The standard also offers several ways of packing the pixels into the image. The basic approach is *sequential coding*, which adds the pixels in the typical left-to-right, top-to-bottom order. A solution known as *progressive encoding* is similar to the interleaving of lines used in the GIF standard. This transmits every nth line at first and then follows with the data in the middle. The result is an image that starts off coarse and is gradually refined as the intermediate data arrive to fill in the gaps. This process is popular on the Net. A third choice is *hierarchical encoding*, which compresses the image at several different resolutions. That is, it breaks it up recursively into subimages and tunes the amount of quantization and bit allocation to the structure of the image. If only a low resolution is necessary, only the lower resolution version can be decoded to save time. Sequential coding is used almost universally right now, but the other versions are becoming more popular as sophistication grows.

The standard also allows the image to be broken up into tiles of smaller JPEG images so that each tile can be compressed with a different level of quality. Ideally, some algorithm would identify the sections that needed high quality and the sections that could survive heavy compression and then apply the appropriate algorithm. Faces, for instance, could be compressed lightly to preserve recognizability, while torsos could be compressed heavily to save space. The greatest problem with this solution is finding an automatic algorithm for identifying tiles. This system is rarely used today but may be adopted more in the future.

One of the more interesting parts of the standard is its flexibility and extensibility. The standards committee is actively exploring ways to add new methods and approaches such as fractal compression or wavelet compression to the mix. The standard is also flexible enough to allow substantial variation in the quantization and Huffman compression used in the most common version. The header of a JPEG file can contain all of the information necessary to decode it, which allows the programmer/user some flexibility in choosing the right values and algorithms.

Some corners of the standard are affected by patents. The arithmetic coding solutions embraced by the committee, for instance, may be covered by patents held by IBM. The details of these patents are often fluid for a number of reasons. The first may be that the patent system is somewhat ambiguous, and the strength of patents is firmly established only through litigation. This step is so expensive that people usually err on the side of caution in order to avoid the problem. The second important reason is that companies often give away patents or issue general licenses for standards in order to encourage use of the technology.

These are just some of the facets of the JPEG standard. More are certain to emerge in the future, because the standards committee continues to entertain additions and revisions in the hope of making the standard more inclusive.

10.2 Basic JPEG

Here are the basic steps of the JPEG algorithm.

1. Convert the data into the right format. Generally, images in the RGB color scheme are converted into YCrCb or YUV. The human eye is very sensitive to slight variations in luminance (brightness) but more forgiving if greater variations in the chrominance (hue). This change lets the JPEG system com-

press the chrominance more to save more space.

2. (Optional) Some versions of JPEG will *downsample* the chrominance values to save more space. This means converting an $m \times n$ grid of pixels into an $\frac{m}{2} \times \frac{n}{2}$ grid by averaging the chrominance of four pixels and converting it into one. Actually, there are several different variations of how this step can be accomplished. Some versions will only downsample horizontally, and two pixels are converted into one. Others will do both. The variations are known colloquially as *2h2v* or *2h1v*. This produces a huge reduction in the image file at the cost of less chromatic variation. Most images never suffer from this change.

JPEG does not need to use 8×8 blocks. This size was chosen because there were many good chip implementations already. In the future, more powerful chips may achieve more compression by using larger blocks.

3. Separate the pixels into 8×8 blocks. This is done for each of the different pixel components separately. If the chrominance values are downsampled, then it is not really correct to say that this step is producing 8×8 blocks of pixels. It is really producing 8×8 blocks of numbers, and these only correspond to pixels directly for luminance values. The chrominance values may correspond to 8×16 or 16×16 blocks depending on how downsampling was done.

   The separation process is normally done in a standard left-to-right, top-to-bottom approach. The progressive and hierarchical approaches require a bit more sophistication because they save some space by sending only the differences.

4. Process each block with a discrete cosine transform. The JPEG standard first subtracts 128 from each value before doing this, because it concentrates many of the values around zero. The DCT converts the 8×8 block of intensity values into an 8×8 block of frequency strengths, that is, coefficients for the various cosine functions in the summation. More data will be saved here by ignoring the high-frequency components, which often make only a difference in the details. The higher-frequency coefficients are often quite small and add only a small amount of shading differences to the image.

5. Quantize the luminance values with a *luminance quantization matrix*. This process is not matrix multiplication or manipulation. The process is simple scalar division and round-off. Let $DCT_{i,j}$ be the result of computing the discrete cosine transform of an 8×8 block of data. Let $QL(x)$ stand for the function that quantizes the luminance values. $QL(x) = round(\frac{DCT_{i,j}}{L_{i,j}})$, where $L_{i,j}$ is the luminance quantization matrix and $round()$ is the rounding

16	11	10	16	24	40	51	61
12	12	14	19	26	58	60	55
14	13	16	24	40	57	69	56
14	17	22	29	51	87	80	62
18	22	37	56	68	109	103	77
24	35	55	64	81	104	113	92
49	64	78	87	103	121	120	101
72	92	95	98	112	100	103	99

Table 10.1: One typical quantization matrix used for luminance values in the JPEG standard.

function that chooses the nearest integer. Table 10.1 shows a commonly used version.

This table was determined through empirical study. The values were made as large as possible to increase compression without degrading the results. The values in the upper left-hand corner are smallest because these are the lower-frequency components of the DCT. Smaller values means that these components are quantized with more precision.

You are free to substitute your own table with a JPEG image and embed the table in the header. While this takes up more space, it can be useful to optimize the values in the header for a particular image. In fact, most high-quality JPEG implementations do this, so there is no real space gain. Fabric samples, for instance, may have much more significant high-frequency components coming out of the DCT, and it might make more sense to keep the precision here.

6. Quantize the chrominance data using the same process but with a different matrix tuned to the lower sensitivity of chrominance information. Table 10.2 shows a common example.

   This example keeps much less precision than the sample luminance quantization matrix defined above. A matrix filled with smaller values might be desirable in situations where absolute color verity was important. This can also be accomplished by substituting a different matrix in the header of the compressed file.

7. At this point, most of the information will be packed into the upper left-hand corner of the 8×8 matrix for two reasons. That is another way of saying

17	18	24	47	99	99	99	99
18	21	26	66	99	99	99	99
24	26	56	99	99	99	99	99
47	66	99	99	99	99	99	99
99	99	99	99	99	99	99	99
99	99	99	99	99	99	99	99
99	99	99	99	99	99	99	99
99	99	99	99	99	99	99	99

Table 10.2: Another typical quantization matrix used for chrominance values in the JPEG standard.

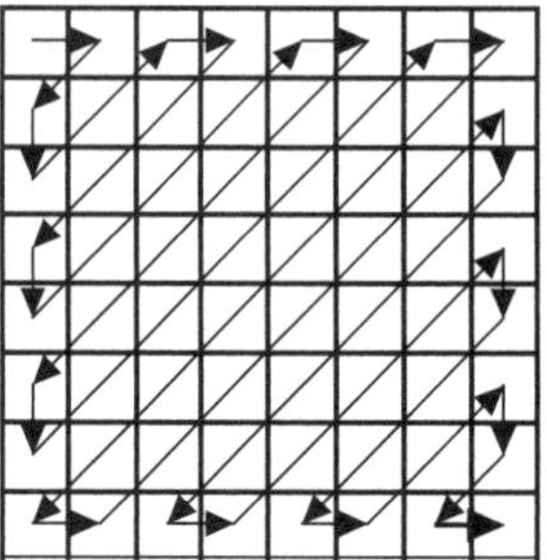

Figure 10.1: The zigzag pattern used to unpack an 8×8 grid of quantized DCT values. The upper left-hand values are unpacked first because they are the most likely to be nonzero and contain valuable information. Run-length encoding allows space to be saved with this unpacking procedure

that the coefficients in the upper left-hand corner are much more likely to be nonzero because of the effects of quantization. First, most images have fairly small coefficients for the higher frequency components. Second, these coefficients often are reduced to zero by the quantization steps. So, the upper left-hand corner of the matrix will have most, if not all, of the nonzero values after quantization.

The JPEG standard committee came up with a zigzag way of unpacking a matrix to take advantage of this system and provide for a way to use run-length encoding to save space.

Figure 10.1 shows the zigzag pattern that begins with the (0,0) value and procedes from there. This packs the values most likely to be nonzero and

Magnitude Range	Luminance Code	Chrominance Code
0	00	00
1	010	01
2...3	011	10
4...7	100	110
8...15	101	1110
16...31	110	11110
32...63	1110	111110
64...127	11110	1111110
128...255	111110	11111110
256...511	1111110	111111110
512...1023	11111110	1111111110
1024...2047	111111110	11111111110

Table 10.3: Examples of the codes used to pack DCT coefficients in the JPEG standard.

ends with the higher-frequency terms that are most often zero. This makes run-length encoding of the zero terms feasible.

8. The value in the upper left-hand corner, $DCT_{0,0}$, is a special value that is most responsible for the look of an 8×8 block because it essentially specifies the average value for the entire block. For this reason, it is encoded with a special *predictive compression* scheme that stores the differences between blocks to save resolution. Imagine that a chain of blocks had $DCT_{0,0}$ values of 319, 345, 400, and 385. These would be stored as the differences: $319, 36, 55$, and -15. The last three terms are often called *predictive errors*. Most adjacent blocks have similar values of $DCT_{0,0}$, so they often have small differences, and this justifies using the predictive scheme. These smaller differences can be expressed in a small space with higher, lossless precision.

9. The predictive error of the $DCT_{0,0}$ for the values is assigned a particular category based on its magnitude. This solution allows small "errors" to be coded in a smaller number of bits. Table 10.3 shows the codes.

   This table was assembled from statistics about the actual variation in the luminance and chrominance in between blocks. It should be clear that differences of between 8 and 32 are much more common in the luminance than in the chrominance. They have shorter codes. Notice that the Huffman tree

The flexible-length codes used for the luminance and chrominance codes are generalized on page 73.

is not completely utilized — there are no codes that are all 1.

The Huffman code from this table indicates the range. The significant bits of the amount of the difference are appended afterwards. For instance, if the luminance differed by six, then the code '100' would indicate the range and the value '110' would indicate the difference six. This produces '100110'. In a similar fashion, the difference of 18 in the luminance would be encoded '11010010'. The size of the range from the first part of the code determines how many bits afterwards are needed to compute the exact difference.

Negative values are stored differently. The table above shows the "magnitude of the differences", which means that it also applies to negative numbers. The table of values is used the same way, but the bits afterwards are encoded as *ones compliment* – essentially the same value with the bits flipped. So a difference of -5 is shown as '110' concatenated with '010', which is the ones-compliment version of 5.

A value of 0 is not followed by any bits.

10. The other 63 luminance and chrominance quantization values are coded up by a version of Huffman coding that analyzes both the run-length and the number of bits in the value. They are *not* coded with difference coding because most of them are usually pretty small anyways.

    They are processed in order specified in Figure 10.1. The Huffman table is keyed upon the run of zero components followed by the size of the nonzero component. So the values $0, 1, 0, 0, 5, 17$ would be encoded by three pairs: (1,1), (2,3), and (0,4). The first value in the pair is the number of zero values and the second is the number of bits nonzero following value.

    A more extensive table is used to convert these pairs into Huffman codes. The actual value is appended afterwards. So, the stream of values $-7, -2, 0, 1, 1, 0, 0, 0, 0, 0, -1$is converted into the pairs: $(0, 3), (0, 2), (1, 1), (0, 1), (5, 1)$. These pairs are looked up in the table and converted into pre-fix codes that designate the number of bits that follow. (0,3) is converted into '100' in the luminance tables. It is followed by the three bits, '000', which indicate -7 in ones-compliment.

    An "end-of-block" code is used to signify the last nonzero code. In most cases, this avoids adding even small bits to indicate the 50 or so zero coefficients that are common in relatively flat 8×8 blocks of pixels.

The section beginning on page 173 describes how to embed additional information in this process. It's useful to hide data that might be used to identify ownership.

11. The values for the block are packed together with the bits for the $DCT_{0,0}$ at the beginning followed by the bits for the other coefficients that run up to

the end-of-block code.

The result is often a fairly small block of bits. It is not uncommon for a fairly flat block to be reduced to the order of 35 to 40 bits, a ratio of about 15:1.

This system is usually known as the *baseline* compression method.

10.3 JPEG Enhancements

A number of different enhancements to the JPEG compression standard are already defined. Researchers are also looking for other ways to extend it in the future. The goals are fairly simple: Increase compression, increase usability, and increase quality. It is not necessary to do all three things at the same time, although people try. The standard is designed to have a very flexible header structure that makes it possible to add many different forms of enhancements over time and have them work successfully. The result is a fairly baroque mixture of compression ideas all linked together under one "standard".

The basic enhancements can be summarized as such:

Progressive Presentation Many photographs travel over slow communications links such as the Internet, and it is often desirable to allow the computer on the other end to display partial images while the rest is arriving. The baseline system will show the pixels arriving as a slow scan running left-to-right, top-to-bottom. A progressive system will transmit a small amount of each pixel, which will allow the receiving computer to display a blocky, pixelated image that is later refined as more information arrives.

Lossless Encoding The baseline method does not recover the image in exactly the same form, so it is often known as lossy. A lossless method makes no mistakes. This is essentially an entirely new compression algorithm that just happens to exist under the rubric of JPEG.

Arithmetic Coding Some systems use arithmetic coding instead of Huffman coding in the last step where the coefficients are converted into the final bit representation. Some claim that this choice leads to files that are about 10% smaller. The problem is that the standard method of arithmetic compression is covered by patents held by IBM, so many people shy away from the approach.

Hierarchical Coding This is slightly different from progressive presentation, but the effect is the same. In this case, the image is first reduced to a number of smaller sizes. So the $m \times n$ image is converted into a $\frac{m}{2} \times \frac{n}{2}$, $\frac{m}{4} \times \frac{n}{4}$, and an $\frac{m}{8} \times \frac{n}{8}$ image by downsampling. The lowest resolution image is shipped first, followed by the differences that are necessary to add detail. The result can be fairly efficient and also allow the receiver to display partial images as the data is transmitted.

Greater Precision The baseline process begins with images encoded with fairly standard eight-bit precision. Better images now come with twelve-bit precision, and it makes sense to modify the JPEG algorithm to take advantage of this.

Variable Quantization The baseline process uses one set of quantization matrices for the entire image. There is no reason why multiple ones cannot be used for different parts. A region with important detail might use a smaller quantization matrix that allows greater detail to pass through while the background may be encoded with a bigger matrix that compresses more details.

Tiling Images Another approach to achieving the same result is creating a tiled version of different images, that is, cutting the image up into rectangles and encoding each rectangle as a different JPEG image. Some could get high resolution and others low resolution. Obviously the amount saved must overcome the greater overhead. Small sections of tiled images can be recovered without decompressing the rest of the image, and this may make the system useful for large images.

Selective Refinement A third approach to saving space in less important regions of the photograph is called "selective refinement", achieved by coding the entire photograph at a low resolution. Then, additional rectangular patches are added to the file, and these patches include extra information that can be used to selectively refine the pixels in this rectangle.

10.4 Lossless JPEG

The lossless version of JPEG replaces the discrete cosine transform with a version of predictive coding that will reproduce the pixels exactly. It would be possible to have the discrete cosine transform reproduce the image exactly by keeping all of the coefficients around in high precision, but it would be a waste of computation. The DCT requires a large amount of computation, and its main purpose is to reduce the amount of information kept around.

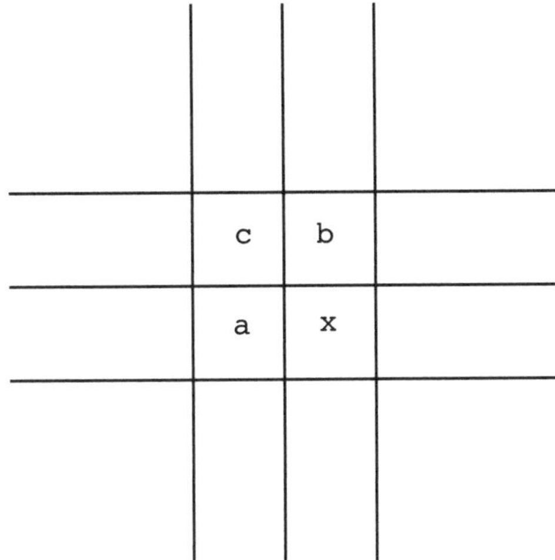

Figure 10.2: If x is the pixel being predicted, then a, b, and c, are the possible choices to use as a basis for the prediction.

The prediction system used in lossless JPEG is fairly clever. It recasts each pixel as the difference between the pixel and an adjacent one. In the parlance of the system, a set of previously encoded pixels *predicts* the pixel, and thus it is only necessary to encode the *error* between this prediction and the real value. In essence, it encodes each pixel as the difference between it and a previous pixel. Usually, these differences are quite small, so it is possible to use many of the range tricks described to encode coefficients in the baseline system. In fact, many of the same tables are often used.

Figure 10.2 shows four pixels. x is the pixel being encoded by the prediction process. $a, b,$ and c are three candidates that were previously encoded and can now be used as the basis for the prediction. Let $P_a, P_b,$ and P_c stand for their values respectively. The lossless version of JPEG uses eight different modes of prediction:

Number	Formula
0	none
1	P_a
2	P_b
3	P_c
4	$P_a + P_b - P_c$
5	$P_a + \frac{P_b - P_c}{2}$
6	$P_b + \frac{P_a - P_c}{2}$
7	$\frac{P_a + P_b}{2}$

The prediction for each pixel is used to calculate the error. This is usually a small number, so the process of packing the differences is important. The scheme is the same as the one used to pack up the coefficients to the DCT in the baseline system. First, the number of significant bits in the error is calculated. These are then converted into packed versions using the codes in Table 10.3.

An error of five in the luminance value is converted into the range code, '100', followed by the three bits representing the error '101' producing the result '100101'. This approach saves a fair number of bits when there are small errors. The best results come when pixel values are duplicated. These are replaced with the code '00'.

10.5 Progressive Transmission

The progressive transmission option for JPEG is designed to make it possible for the image to be reconstructed in several different stages. This is useful when the image is transported over a slow medium, because it allows the recepient to view a rough version of the image before the entire file arrives.

The progressive option accomplishes this task by shipping the coefficients from the 8×8 block in seven different stages. Figure 10.3 numbers the coefficient with the stage. The coefficients in the upper left-hand corner travel first because they offer the most basic information about the image. The other coefficients often some detail, but they are often quantized to zero to save space.

The upper left-hand coefficient is the average value of the entire 8×8 block of pixels. It makes a good surrogate for the entire image. The block is refined by each successive stage. The additional information that travels in each successive stage is much like the additional information that is packed into a JPEG file as the image is encoded at a higher quality level. The lowest quality level uses higher quantization levels that zero out the coefficients in the later stages. The higher quality levels use lower quantization levels that keep more information in the later

1	2	3	4	5	6	7	
2	3	4	5	6	7		
3	4	4	5	6	7		
4	5	5	6	6	7		
5	6	6	6	7			
6	7	7	7				
7							

Figure 10.3: This grid divides the 8×8 grid of discrete cosine transform coefficients into eight different stages. The first seven stages are numbered. The remaining coefficients, which are often zero, are sent in the eighth stage.

rounds.

Some other JPEG users are experimenting with sending the bits in successive stages as well. This means that the most significant bits of the coefficients are shipped in the first round. The less significant bits follow in successive rounds or stages.

Here's an example built around the following chain of coefficients: 1031, 265, 24, 9, 3, 2, 2,1. These values are '10000000111', '100001001', '11000', '1001', '11', '10', '10', '1' in base two. The values can be shipped in three stages. The first contain the first four out of twelve significant bits; the second contains the next four; and the third contains the last four. Only the first two values have any bits that will be shipped in the first stage. They are 100 and 1. The second stage contains three values: 0000,0000, and 1. Only the third stage needs to include bits from all of the values: 0111, 1001, 1000, 1001, 11, 10, 10, and 1.

There are several different varieties of JPEG out there that use various combinations of these two schemes to do progressive transmission of JPEG images.

10.6 Hierarchical Transmission

The goal of hierarchical transmission is similar to the goal of progressive transmission: Repackage the image at lower resolutions to handle limitations in the display hardware. In this case, the goal is accomplished by storing the image at several levels of lower resolution. A lower-resolution screen will not need to unpack the entire image and downscale it to display it. It can get by with one of the smaller images transmitted at the beginning of the file. If the user wants to zoom in on one section, then the higher resolution image can be decoded.

There is one basic step used to compress an image for hierarchical transmission. It takes an $m \times n$ image and creates an $\frac{m}{2} \times \frac{n}{2}$ image by downsampling. This step can be repeated multiple times to create as many images at successively lower resolutions as may be desired. Shipping the extra images need not take more space, because the higher-resolution image is not shipped. It is just represented as the error in the prediction of the lower-resolution image.

Here are the steps to creating one lower resolution image. Repeat as necessary.

1. Begin with image I_i, which is $m \times n$. This process will create I_{i+1}, which is $\frac{m}{2} \times \frac{n}{2}$ pixels. It will represent I_i as the error produced by the prediction provided by I_{i+1}.

2. Create I_{i+1} by downsampling the image. This can be accomplished by averaging the four pixels. A more sophisticated approach takes antialiasing into account and averages in other adjacent pixels. This sophisticated approach can be a distant weighted average of any combination of nearby pixels. For practical purposes, it is usually just the eight additional adjacent pixels.

3. Now represent I_i as the error after re-expanding I_{i+1} and using it to predict the values of I_i. I_{i+1} can be re-expanded by simply converting one pixel into four. A better approach is to use bilinear interpolation and set the four pixels to be weighted averages.

4. Use the DCT on the error terms. These should be relatively small and should be quantized with a smaller matrix to preserve the higher detail. Many should be zero.

5. Pack these DCT coefficients as before.

This process begins with the original image set to be I_0 and can be repeated as many different times as necessary. The entire package of images at different resolutions does not need to be substantially larger. In fact, many of the coefficients of the error terms are zero, and the files can often be close to the same size.

10.7 Conclusions

The JPEG format is now a well-known format for image compression that is widely used on the Internet. It achieves its great compressive strength by using a discrete cosine transform to model the frequencies of each 8×8 block of pixels. The most significant frequencies are kept while the least important ones are reduced to zero. The process of keeping and throwing away coefficients for the DCT is where much of the art of JPEG compression comes into play. The JPEG standard does not specify how the decisions are made and leaves this up to each compression manufacturer. As a result, there can be great differences between the quality of one implementation and another.

The JPEG standard is generally a lossy standard. While there is a lossless version, most people use the standard's ability to throw away needless information to save space.

This chapter has also mentioned a number of different options that can be used in implementations of JPEG images. While the JPEG committee has developed and implemented these ideas, they are not widely deployed in commercial applications. This is largely because the greatest demand for JPEG compression is for distributing images over the Net. If other uses become more common, then the commercial solutions will probably follow.

The MPEG standard in the next chapter is similar in style to the JPEG standard described here. It also uses the DCT to compress 8×8 blocks of pixels, but it includes several important features for saving space when the sequences of moving images repeat many details.

Chapter 11

Video Compression

Digital images are another area begging for compression algorithms. Moving pictures are a popular medium for expression, but the amount of information is staggering even after the relentless success of the magnetic storage industry. Good video signals change the image 20 to 30 times per second, which means that even a minute of video is made up of 1,200 still images.

The good news is that there is plenty of redundant information in the images that can be excluded to compress the file. More importantly, the eye is often quite forgiving, so the reconstruction does not need to be perfect. The video algorithms are all lossy, and this allows them to save substantially more space.

The video algorithms are all close cousins to the JPEG standard described in Chapter 10, because they use the Discrete Cosine Transform to reduce 8×8 blocks of pixels. The coefficients are also quantized and packed with a Huffman-like compression scheme to reduce the size. The major difference is that an addition step known as *motion estimation* is used to squeeze out difference in moving frames. In many cases, large parts of the image remain unchanged between frames. In other case, large parts only move over a pixel or two. The motion estimation looks for these cases and avoids transmitting the same information twice. This motion estimation is responsible for much of the compression achieved.

Discrete Cosine Transforms are a type of wavelet compression described in Chapter 9.

There are a number of different video compression standards. Most of them are variations of the work of two different committees: the International Telephony Union's Specialists Group on Coding for Visual Telephony and the International Standards Organization's motion picture expert group, known as MPEG. All adhere to the basic framework dictated by merging motion estimation with discrete cosine transforms, but each standard is tuned to different needs.

Some cameras use Motion-JPEG, which simply applies JPEG to each frame without doing any interframe analysis.

In general, the MPEG standard is more complicated because it is designed to support more sophisticated multimedia presentations and motion pictures. The H.261 standard, which can reduce a video signal to 64kbs, is aimed more at low-bit-rate video telephony. It is now being superceded by the more sophisticated revision known as H.263. Television broadcasters and motion picture packagers who use MPEG, which can use up to 1.8mbs, are less forgiving of glitches and errors than the average video telephone call user. The compressed digital video must compete with fairly successful analog technology. Video telephony has no competition so it can get by with much glitchier quality.

The fractal coding systems developed by Iterated Systems are the only major approach to image coding that is signficantly different than the wavelet methods exemplified by discrete cosine transform. Chapter 13 describes it.

The most popular standards are the ISO's MPEG-1 and the ITU's H.261, but others are frequently emerging as groups tune the different standards for their use. The MPEG standard, for instance, has been revised twice, producing the MPEG-2 and MPEG-4 standards. At this writing, MPEG-7 is in development. There are also many other standards used in different situations. Some are aimed at producing a very small size for use in low-bandwidth applications such as digital video phones. Others aim for very good reproduction to provide perfect images for theatres. This book cannot cover all of the different standards because of space, so it concentrates on the major two. Most other solutions are variations on this theme.

11.1 Pixel Details

Chapter 10 describes the JPEG standard.

At the most basic level, the moving pictures are made up of colored pixels,m and these colored pixels are compressed by grouping them into 8×8 pixel blocks that are compressed with discrete cosine transforms. This part of the structure is similar to JPEG. Video compression, however, requires a much richer structure to support the motion estimation that generates most of the compression. The pixels are placed into a more complex hierarchy that needs definition. This is because the MPEG standard wants to be as compatible as possible with the older analog television formats.

The video compression standards are built around the four basic common analog television standards: NTSC, PAL, and SECAM. In order to simplify the process, the standards committees developed common formats that can interact with all of the analog standards. H.261 uses the Common Intermediate Format (CIF) and the Quarter Common Intermediate Format (QCIF), which takes one quarter of the space. The MPEG committee chose the term *Source Input Format* (SIF) to do the job. It comes in two sizes: SIF-525, which is used with NTSC video, and SIF-625 for PAL video.

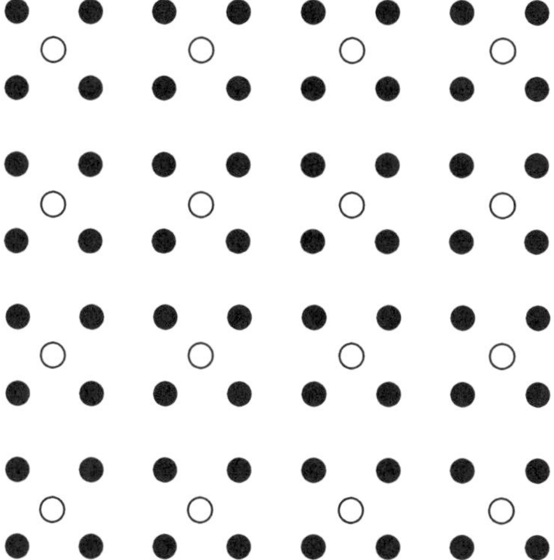

Figure 11.1: The location of the luminence (black) pixels and the chrominance pixels (white) in 4:2:0 mode of H.261.

The following table summarizes some of the salient information about the formats:

	CIF	QCIF	SIF-525	SIF 625
Luminance (Y)	360×288	180×144	360×240	360×288
Chrominance (CR and Cb)	180×144	90×72	180×120	180×144
Frame Rate	29.97	29.97	30	25

Figure 11.1 shows how the chrominance pixels are interleaved with the luminance pixels.

CIF and QCIF for H.261

The Common Intermediate Format (CIF) and its quartersized companion the QCIF are designed to provide enough structure to support robust motion estimation. The first step in this process is to strip off some pixels from the left and right sides of the image. Four pixels are removed from each side of the CIF leaving a 352×288 pixel image, and two pixels are removed from each side of the QCIF, leaving a 176×144 image. The result is called the *significant pel area (SPA)*, and the two strips give the image in the SPA room to move.

The pels in the SPA are aggregated into a four-level hiearchy:

Block Layer The pels are grouped into 8×8 blocks that will be fed into the DCT.

Macroblock Layer Four blocks form a macroblock. This is a 16×16 block of luminance pels. The chrominance information is usually 8×8.

Group of Blocks Layer Thirty-three macroblocks make one "group of blocks" (GOB). These are arranged in three rows of eleven macroblocks.

Picture Layer There are twelve GOBs in a CIF picture and three in a QCIF. These are arranged in six rows of two GOBs in the CIF and three rows of one GOB in the QCIF.

Figure 11.2 illustrates these four layers.

MPEG's SIF Structure

The MPEG uses a similar structure for all of its compression. The generic video formats such as NTSC or PAL are converted into the SIF by any one of a variety of decimation techniques that average pixels together. The SIF allows a flexible number of lines and pixels per line, but these are all grouped together as 8×8 blocks and 16×16 macroblocks of luminance data. Each macroblock also comes with two 8×8 blocks of chrominance data. There are no strips of pixels removed from the edges, as there are in H.261.

The MPEG algorithm groups the macroblocks into *slices*, which are just single rows of macroblocks that are subsequently grouped together to make a picture.

The MPEG algorithm also contains a larger abstraction known as the *group of pictures*. There is no maximum (or minimum) size of a group.

11.2 Motion Estimation

The process of motion estimation is responsible for a large amount of the compression achieved by these schemes. The goal is to try to represent each block of pixels by a vector showing how much that block has moved between images. This is particular useful for the background of images because this fixed scenery can be replaced by a fixed vector of (0,0). The process is normally only done for luminance values.

The vector also makes it possible to save plenty of space when parts of images are panning across the screen. If the camera is moving gently, then most of the

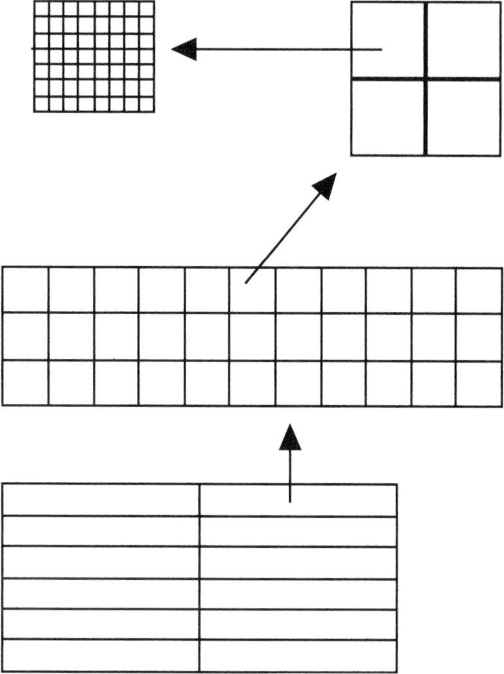

Figure 11.2: The four hierarchies defined for the CIF. From left to right, top to bottom: the block, the macroblock, the group of blocks, and the image itself.

blocks will move a fixed amount, say two pixels to the right and three pixels up. These blocks of pixels are usually replicated exactly when there is scenery in the background.

The motion vector can also save space even when the block of pixels is not an exact duplicate of the other pixels. In this case, the difference between the two can also be coded with a DCT. The differences are usually quite small, and this reduces the size of the coefficients and thus the number of bits necessary to encode the changes. For instance, someone might be throwing a ball across the screen. The motion estimation can compress most of the scene, while the DCT of the difference can handle the gradual darkening.

In these two major algorithms, the motion estimation is usually done on the macroblock level. The H.261 algorithm uses simple *forward prediction*, which computes the difference from only a previous frame that was already decoded. This is fairly simple to implement, and it is also fairly easy to run in real time. The MPEG algorithm, on the other hand, includes *backward prediction* and a combination of the two known as *bidirectional prediction*. This means that sometimes the information about the pixels in a frame depends on both the last frame just decoded and the frame about to follow. This does not introduce a logical loop that crashes the machine because the MPEG standard specifies that complete frames must be inserted every so often to minimize errors and ensure that the display can resynchronize itself if several frames disappear. The prediction can be worked backward from these fixed images. The frames themselves are stored in the order they will need to be decoded, not the order they will be displayed.

The MPEG uses backward prediction because it helps save space when information is being uncovered. Imagine that the video is of a curtain unveiling a scene. The complete area behind the curtain will emerge eventually. The new information emerging at the edge of the curtain can be copied from the image that contains everything behind the curtain. The backward prediction makes it possible to save space by going forward and copying this information from a future frame. Naturally, this process adds significantly more computation to both the compression and the decompression process.

The process of calculating motion estimation is time consuming, and people have developed a number of different shortcuts to save time. The basic difference is measured with several metrics. The most common one is the mean absolute error, which is also called the L_1 norm. If A and B are two images, then

$$||A - B||_1 = \sum_i \sum_j |A_{i,j} - B_{i,j}|.$$

The L_2 norm is

$$||A - B||_2 = \sum_i \sum_j (A_{i,j} - B_{i,j})^2.$$

The L_1 norm is usually employed because it is cheaper to calculate, but newer hardware is reducing the practical difference. The extra penalty paid for deviation is not important, because the algorithm is looking for the closest match, and this is usually fairly small.

The goal for the MPEG and H.261 encoder is to take each 16×16 macroblock and try to find an offset of (l, m) pixels that minimizes the L_1 norm. Ideally, this minimum will be zero because the 256 luminance values and 128 chrominance values are exactly the same. The best algorithm is a brute-force approach that will try all possible values of (l, m). This approach isn't particularily practical, so some compression tools use shortcuts that work fairly well.

The shortcuts for motion estimation check a few sample points and then use the results to guide the search for more sample points. Here's one three-step approach [RH96] :

1. Start with $(l, m) = (0, 0)$.
2. Compute the L_1 difference at (l, m), $(l + 4, m)$, $(l, m + 4)$, $(l - 4, m)$, $(l, m - 4)$, $(l + 4, m + 4)$, $(l - 4, m + 4)$, $(l - 4, m + 4)$, and $(l - 4, m - 4)$. Find the minimum of these nine offsets.
3. Set (l, m) to the minimum.
4. Compute the L_1 difference at (l, m), $(l + 2, m)$, $(l, m + 2)$, $(l - 2, m)$, $(l, m - 2)$, $(l + 2, m + 2)$, $(l - 2, m + 2)$, $(l - 2, m + 2)$, and $(l - 2, m - 2)$. Find the minimum of these nine offsets.
5. Set (l, m) to the minimum.
6. Compute the L_1 difference at (l, m), $(l + 1, m)$, $(l, m + 1)$, $(l - 1, m)$, $(l, m - 1)$, $(l + 1, m + 1)$, $(l - 1, m + 1)$, $(l - 1, m + 1)$, and $(l - 1, m - 1)$. Find the minimum of these nine offsets.
7. Set (l, m) to the minimum. This is the offset.

The MPEG algorithm may use half-pel accuracy. In this case, the three steps can be turned into a four-step algorithm by adding an additional round to reduce the accuracy even more.

Some implementations save space by subsampling the 16×16 block and doing the compression on the subsampled version.

MPEG's Prediction Types

The frames of the H.261 can use motion estimation to predict the value of the next frame. That is, the information in one frame serves as the basis for the construction of the next one. The MPEG system uses a more complicated arrangement in order to save more space and make it easier to recover if crucial frames disappear. There are four types of frames in the MPEG world:

I Frames These are complete frames including all of the information necessary to reconstruct them. They do not depend on any future or past frames for their display, and they can be extracted from the video stream and displayed on their own. These frames serve as reference points that are included to reduce errors and allow the compression scheme to recover if glitches drop a frame or two.

P Frames These are *forward prediction* frames that use a previous frame to define their structure. This basic format can also be used as the basis for other P frames, which means that errors can accumulate if several P frames are used in a row.

B Frames These are mixtures of both forward and backward prediction intended to maximize the amount of compression. These frames can't be used as the basis for predicting other frames in order to reduce the amount of error accumulating over sequences. They must be based on either P or I frames that occur both before and/or after it.

D frames Each 8×8 block of pixels is reduced to one DCT coefficient, the first one that specifies the average color of the block. Naturally, this produces very small frames, but does it so at the cost of much loss of precision.

The MPEG standard allows the encoding software to interleave the I, P, B, and D frames as it chooses subject to a few simple rules. This allows the encoder to make subjective decisions about quality versus compression size. If the software is sacrificing quality for size, it might include a large number of P and B frames and not worry about the effects of subjective degradation. The biggest problem with B frames is that they spread out the distance between the I or P frames used as references, and this means greater differences between the references and thus more interpolation error.

The choice of the numbers of I, P, B, and D frames has trade-offs, however, because of the fixed bit rate of the MPEG standard. If more I frames are used, then it is often not possible to pack them into the desired bit rate without increasing the quantization size and degrading the images. So adding more I frames may seem to add more quality, because an I frame contains a complete description of the image, but this effect can be ruined by the fact that individual I frames only contain a few of the coefficients necessary to reconstruct each block.

The presence of B frames also complicates the decoding process. These frames rely on both past and future frames, which means that the decoder must know the future in order to reconstruct them. The MPEG algorithm allows the frames to be decoded out of display order so that this can be accomplished. For instance, imagine that the sequence of frames took this pattern: I B B B B P B B B B I. The first I frame can be decoded without a problem because it contains all of the information about itself. The next four B frames, however, can't be decoded without the sixth frame, a P frame, because they were encoded with both forward and backward motion estimation. Also, the seventh through tenth frames can't be decoded without the final I frame. So, the decoding process will work like this: I P B B B B I B B B B. The frames will be reshuffled when they're finally displayed.

Use of Motion Estimation in H.261

The H.261 uses motion estimation if it is desirable. The standard does not force encoders to use motion estimation at all, but most do so because it saves space. Decoders, of course, must be ready to accept motion estimation if they want to process all possible streams of data. The standard also allows each encoder to make a decision on a macroblock-by-macroblock basis. An encoder might use motion estimation for one macroblock if the error is small and not use it for another because the error would be unacceptable. Each encoder is free to make this decision as it wants. In practice, most use fairly simple threshold analysis. If the mean squared error is greater than one value, then the macroblock is encoded with traditional ways. If the error is less, then a motion vector is sent along.

The motion vectors in H.261 can vary between -15 and 15 pixels in both the horizontal and vertical direction. They are calculated using the 16×16 macroblock of luminance values. The displacement for the chrominance values that make up an 8×8 block is calculated by dividing by two.

Mixing motion-estimated blocks with non-motion-estimated blocks can introduce some blocky effects. Some H.261 decoders include an optional set of filters known as *loop filters* that smooth the edges between the macroblocks. These are usually 3×3 filter kernels that do basic averaging.

Use of Motion Estimation in MPEG

The MPEG algorithm also uses a heuristic to determine when it wants to use motion estimation and when it wants to include the raw data about the block itself. The decision varies from encoder to encoder, but the basic strategy is to evaluate the error produced by motion estimation and determine whether it falls under an acceptable level.

The motion estimation process in MPEG is more complicated than the process in H.261 because there may be both forward and reverse estimation. The P frames offer only forward estimation, so they are created in a way that is fairly similar to H.261: The motion estimator looks at the previous image and determines whether there is a similar block that lies within $(\pm 15, \pm 15)$ units away.

The B frames can include forward and/or reverse motion estimation. If only one is around, then it is represented by one displacement vector. If both are used, then two displacement vectors are around. The final result is calculated by averaging both macroblocks together to produce the final result. In general, the encoder will try all three possible solutions and choose the best one if it falls under the error threshold.

11.3 Quantization and Bit Packing

Both MPEG and H.261 use schemes for quantizing the coefficients and packing them with variable length codes (VLC) that are reminiscent of the techniques used with JPEG. (See Chapter 10.) The goal is to find the right step size for the quantizer. Smaller steps mean greater detail but more bits per coefficient. Both systems allow the quantization step size to be adjusted between macroblocks. Both systems also don't specify how the quantization is to be done. They specify only how the decoder will work. It is up for the coder designer to choose the best solution.

H.261 Quantization and Coding

The H.261 system codes the coefficients from the discrete cosine transform in two different ways. The "first coefficient", $DCT_{0,0}$, is coded as the difference between it and the previous macro block. This amount is quantized by a fixed amount, 8. That is, it is divided by eight and rounded down. This quantization simply finds the average value of the pixel because the $DCT_{0,0}$ is eight times the average value.

The 63 other coefficients are quantized by a flexible scale that produces 255 different levels between -127 and 127. The absolute range of the coefficients is fixed to be between -2048 and 2047. If s is the stepsize, then the different levels are:

$$\{255s+1, \ldots, -7s+1, -5s+1, -3s+1, 0, 3s-1, 5s-1, 7s-1, \ldots, 255s-1\}$$

when s is even and

$$\{255s, \ldots, -7s, -5s, -3s, 0, 3s, 5s, 7s, \ldots, 255s\}$$

when s is odd. These levels represent quanta numbered between -127 and 127. The coefficients are capped to range between -2048 and 2047, so in some cases the largest quanta are not used. For instance if $s = 31$, there are only 122 different quanta between -66 and 66. The rest of the levels are set to be either -2048 or 2047.

There are 32 stepsizes: $0 \leq s \leq 31$. The coder can change the level between macroblocks as necessary. In some cases, the coder may look at the outgoing buffer of data and determine that it is close to full. In these cases, the coder might increase the stepsize to zero out more coefficients and save bandwidth. This is the main way that the H.261 adapts to the channel.

After the 64 coefficients are quantized, they must be packed into their final representation. The coefficients are ordered with the standard zigzag pattern used in most DCT systems. (See Figure 10.1.) The sequence is then converted into pairs of numbers signifying the length of zeros followed by the nonzero level. So $15, 2, 0, 0, -3, 0, 1, 1$ would turn into $(0, 15), (0, 2), (2, -3), (1, 1)$, and $(0, 1)$. These would then be packed up using a Huffman table.

The H.261 standard uses a fairly small Huffman tree that only contains a few of the most popular pairs of values. The rest are encoded with a twenty-bit escape sequence composed of six bits of tag signifying an escape, six bits representing the length of the run of zeros and eight bits denoting the level between -127 and 127.

MPEG Quantization and Bit Packing

The MPEG system uses a system for quantizing the individual coefficients that is closer to JPEG than H.261. This means that it has a quantization matrix that contains different quantization step sizes for each coefficient. The H.261 has a uniform stepsize for all coefficients except $DCT_{0,0}$, which makes it easy to change on the fly. The MPEG algorithm also allows the encoder to change the quantization matrix, but obviously it takes more space to ship along an 8×8 grid of values.

Here's the default quantization grid:

8	16	19	22	26	27	29	34
16	16	22	24	27	29	34	37
19	22	26	27	29	34	34	38
22	22	26	27	29	34	34	40
22	26	27	29	32	35	40	48
26	27	29	32	35	40	48	58
26	27	29	34	38	46	56	69
27	29	35	38	46	56	69	83

The first coefficient is coded as the difference between it and the previous first coefficent. What is the "previous" is not immediately obvious. Each macroblock contains four luminance 8×8 blocks and two chrominance 8×8 blocks. Call the luminance blocks a, b, c, d and the chrominance blocks e and f. Normally, a is the predictor for b, b the predictor for c, and c the predictor for d. The $DCT_{0,0}$ coefficient of the last luminance block in one macroblock, d, is the predictor for the $DCT_{0,0}$ of the first luminance in the next macroblock, a. The chrominance blocks are ordered along the scanline or slice as you might expect.

The differential between the current $DCT_{0,0}$ and the previous one is packed with a system very similar to the JPEG solution. First, a Huffman-like code represents the size of the difference and this is followed by the difference itself.

The flexible-length codes used for the luminance and chrominance codes are generalized on page 73.

Range	Luminance Codes	Chrominance Codes
0	100	00
$-1, 1$	00	01
$-3, -2, 2, 3$	01	10
$-7 \ldots -4, 4 \ldots 7$	101	110
$-15 \ldots -8, 8 \ldots 15$	110	1110
$-31 \ldots -16, 16 \ldots 31$	1110	11110
$-63 \ldots -32, 32 \ldots 63$	11110	111110
$-127 \ldots -64, 64 \ldots 127$	111110	1111110
$-255 \ldots -128, 128 \ldots 255$	1111110	11111110

This code is followed by the smallest number of bits necessary to represent the difference. So a value of 5 in the luminance becomes 101101, and -8 becomes 110011.

The other coefficients are quantized with a flexible system that uses the current quantization matrix and a scaling value that can be adjusted throughout the compression process. This is similar to the scale value used by H.261. If x is the value to be quantized, $Q_{i,j}$ is the corresponding matrix quantization step size for

$DCT_{i,j}$, and s is the scale, then the quantization level is

$$\frac{8x}{sQ_{i,j}}.$$

The quantized coefficients are then unwound with the standard zigzag pattern (see Figure 10.1) and coded with a Huffman-like table for run length and level pairs as in JPEG and H.261. The table for MPEG is more extensive than H.261, but it is still missing some pairs. There is an escape code to handle these cases.

11.4 MPEG-2

MPEG-2 is a newer standard developed by the Motion Picture Experts Group designed to revise and extend the MPEG-1 format. Some of the enhancements add more precision to increase the quality of the images, while others provide more places where greater compression can be achieved. All of the enhancements can be parameterized so the compression software can make many different decisions about which features to use and which to ignore.

The biggest enhancement may the be addition of greater detail. MPEG-2 movies can take up to 100Mbs to encode, an amount that is significantly more than the 1.8mbs limit of MPEG-1. This allows larger format images to be encoded in greater detail in order to expand the attractiveness of the format. The system can handle inputs in some of the major HDTV formats and also deal correctly with interlaced formats. Both expand the use of the system into other arenas.

The audio format of MPEG-2 is also extended.

The audio component for the MPEG standards is covered in Chapter 12.

11.5 Conclusions

The MPEG and H.261 algorithms described in this chapter are some of the more important new algorithms, because they allow digital computers to conquer the old analog territory of video images. The traditional digital solution of simply converting everything into an integer fails in this case because video produces such a flood of data that it still can't be readily stored. Only compression can make it manageable.

The algorithms are notable because they extend the discrete cosine transforms described in Chapter 9 and used in the JPEG standard described in Chapter 10. The solution is using motion estimation and cross-frame interpolation to significantly reduce the amount of information being stored.

It is not possible to cover all of the different video formats in this book, but it is possible to give a basic understanding for how the systems work. Here's a basic summary:

Standardize the Format There are many different raw video formats available. All systems convert them into a standard format.

Break the Data into 8 by 8 blocks of Pixels All major systems work on 8×8 blocks of pixels. There is no reason why larger blocks might not be used, but they increase the computational complexity of the algorithm. It is also not clear that they aid compression signficantly. The best compression occurs when an entire block contains close to the same pattern that can easily be represented by one of the 64 different DCT basis functions shown in Figure 9.4. This allows most of the DCT coefficients to be thrown out to save space. Smaller blocks are more likely to be more consistent. Larger blocks offer the chance to save more space at the risk that coherence will be destroyed. 8×8 blocks are a good compromise. If the raw video feeds begin coming in greater resolution, then larger blocks might make sense.

Compare to Other Blocks In many cases, one block is very similar if not the same as another nearby block in an adjacent frame. A significant amount of space can be saved by encoding this block as a vector difference from the reference. If the block hasn't moved at all, then the vector (0,0) is a pretty small representation of an 8×8 block.

Choose Important Coefficients After the DCT of the block is computed, the hard decision is which coefficients to keep and which to exclude. None of the major standards offers any required way to do this. They merely specify the structure for how the coefficients will be transmitted to the decoder. The decisions are often artistic ones, and leaving the rationale out of the standard creates the opportunity for multiple vendors to come up with different solutions that might perform in different ways.

Scale the Coefficients for Quantization The DCT coefficients normally arrive as many-bit, double-precision, floating-point numbers. This detail is rarely necessary. Through a mixture of experimentation and science, each video compression scheme comes with set of standard quantization sizes for each coefficient. While these are replaceable, most systems use the standards. The most important coefficients are quantized in the greatest detail.

Pack the Coefficients Most schemes use a mixture of Huffmanlike entropy coding and run-length encoding to pack the nonzero coefficients into a consistent form. Often, error-detection systems are included to alert the decoder to any wrong blocks.

Chapter 12

Audio Compression

Sound is another rich arena for compression technology. Music is a common commodity sold on digital plastic disks by the recording industry, and voice may still occupy the largest amount of bandwidth carried by the digital networks. This means that good audio compression is important for two major industries, and this means that there has been plenty of research in the area.

The bulk of the research involves identifying the salient frequencies in the signal, keeping the important ones, and getting rid of the others.

See Chapter 8 for information on quantization.

Of course one person's important frequency is another person's noise. There is a fair degree of artistry in the process of designing these compression schemes. This level of artistry even includes situations where the compression is lossless. There is a significant segment of the population who disparage the digital recordings that are commonly circulated as compact disks. They feel that the process of digitization excludes an important part of the musical experience. The process of digitizing the signal loses enough information to make a difference to their ears. The common complaints are that the CD music sounds too crisp and cold compared to the warmer analog recordings circulated on vinyl grooves.

This book is not an artistic treatise and it is not possible for this author to summarize all of the artistic criteria used by listeners. Readers and their ears are on their own in this arena. This is often the case in the real world where some studios sample and revise the parameters of the compression algorithm until a satisfiable version is found.

In many cases, the decision about which frequencies are important is not made by an artist. The phone companies have often excluded the frequencies that aren't important to voice communication. This may hurt the feelings of people who want to play music over their answering machines, but it can lead to a significantly smaller signal. These models for compression rely upon psychoacoustic models developed by audiologists and speech pathologists who study how the brain responds to voice.

But in other cases the artist or recording specialist has some control over how the frequencies are treated. Many of the most popular formats are deliberately vague about the compression process. They specify only the final format for the data. The job of sorting through the frequencies and determining which ones are important is left to the software creator, and each creator takes his or her own path. The result is a wide range of audio compression tools with different behaviors. An artist or recording technician can choose between them to find the best result for each particular file.

In general, the trade-off in speech coding is the standard one: More bits equal better quality. This trade-off is fairly subjective, however, and different algorithms offer different performance at different bit rates. Low bit-rate systems, for instance, are use approaches classified as *parametric coders*, which use some model of the human vocal tract and encode some speech by finding the right parameters from this model to match the speech. These often perform reasonably well between 2.5 and 5 kps, but they transmit only those frequencies used in spoken communication (i.e. 200 to 4000Hz).

The book edited by W. Bastiaan Kleijn and Kuldip K. Paliwal provides a good survey of modern speech coding methods.

At high bit rates, the quality of these parametric methods is usually surpassed by *waveform-approximating solutions* that use the standard frequency analysis to describe the signal. At the rates above 5.0 kbs, the quantization systems can encode enough information to make this workable. In these cases, the systems capture more of the individual idiosyncrasies of the speaker because they don't compress the signal by forcing it to conform to some generic model.

Many of the newer cellular phone systems allow variable-rate encoding so the systems can adjust the amount of bits per second, depending on the call volume. At peak hours, the signal may sound terrible, but at off-peak times it can be significantly better.

Music is coded by an entirely different set of algorithms. These generally use a much higher bit rate and often attempt to capture a much wider range of frequencies (i.e. 20 to 20,000Hz). There is still plenty of effort made at excluding frequencies and sound, and so many people have strong opinions about the success or failure of different encoding systems. Often, these opinions rest on each person's hearing ability. Some people never hear some sounds, and so they don't miss them. Others do, and the loss of them hurts the signal.

In general, the long-distance phone service in the United States began by transmitting only the range between 300 and 3400Hz. There have been some recent efforts to expand this range to increase quality. Lower frequencies add bass resonance that makes voices sound more human. Higher frequencies add definition and help distinguish similar sounds like the 'p' and 'b' or the 'f' and the 's'.

This book cannot do a good job surveying all of the different methods for compressing sounds. The area of speech compression has been an active arena for telephone companies for some time. They have many different variations of the different encoders, and the best books on the subject focus on speech alone. Audio encoding is a more recent development because music and multimedia production went digital after the telephone network. The book will try to cover several of the major standards in order to make it easier for users to get a feel for the systems.

12.1 Digitization

Audio signals can be recorded in many ways, but this book is concerned only with digital versions. The basic scheme for digitizing a signal is known as *pulse code modulation*. That is, the intensity of the signal is taken n times per second and turned into a number. The more times the signal is sampled, the higher the frequency that is stored. This value, $\frac{n}{2}$, is known as the *Nyquist frequency*, which is the highest frequency that can be represented with n samples per second.

Most humans can't hear frequencies greater than about 22,000 Hz, and so the digitization process used to create compact disk audio samples 44,100 times per second. This produces a stream of numbers that must be represented in digital form, and one of the major choices is the number of bits in these numbers. Sixteen bits offer a nice amount of resolution, but the files can be large. Eight bits don't offer enough resolution, and the size of the steps between the 256 possible values is large enough to add a fair amount of noise to the signal. Twelve bits per value would be enough for most cases, but it is difficult for computer architectures to handle because it is an odd size. The files are also a bit big.

One of the better solutions is known as μ-law encoding, and it uses a tiny version of scientific notation to store the size in eight bits. The first bit is given over to the sign, the next three are used for the exponent of the value and the last four hold the log of the value used as the mantissa. This isn't very precise, but it concentrates the precision at the right levels. The small values have a small step in between, while the larger ones have a larger step size. This tuned quantization minimizes the "noisiness" of a signal.

Here is the central part of the source code provided in the `comp.speech` FAQ. The code is attributed to Craig Reese and Joe Campbell.

```
sign = (sample >> 8) & 0x80;
exponent = explut[(sample >> 7) & 0xFF]; mantissa
= (sample >> (exponent + 3)) & 0x0F; return(
~(sign | (exponent << 4) | mantissa) );
```

In this case, the array `explut` is a 256-entry array used to look up the log of the number. 0 and 1 return 0; 2 and 3 return 1; 4, 5, 6, and 7 return 2; 8 through 15 return 3 etc.

The μ-law encoding is a fairly effective compromise for storing data, but it is still not too efficient. In fact, many people don't consider this to be data compression. At 44,100 samples per channel per second, then it's using 352kbs for just one channel even after μ-law encoding.

Predictive Huffman Compression

One of the basic solutions to compressing pulse-code modulated data is to construct a table of differences between the individual samples and use Huffman compression to encode them. In most cases, the differences are small, because the signal strength does not change markedly between samples. This is especially true at high-quality sampling rates used for CDs.

This system can easily reduce the size of a file by a factor of two or three if the signal is relatively simple and dominated by lower frequencies. Higher frequencies reduce the effectiveness of this approach, because the file values are more likely to change wildly from sample to sample.

12.2 Subband Coding

One of the biggest techniques used in audio coding is known as *subband coding*, because it revolves around analyzing a signal to find its component frequencies and then allocating these frequencies to various subbands, which are compressed separately. That is, a signal might be analyzed to find the intensity of the frequencies between 0 and 400Hz, 401Hz and 1200Hz, 1201Hz and 1800Hz, 1801Hz and 3000Hz, and 3001Hz to 20,000Hz. Then, these five subbands are analyzed separately.

One of the reasons that subband coding works well is a phenomenon known as *masking*. While the ear may pick up all frequencies in a signal, the brain throws out many of them in the process of actually "hearing" or taking apart the signal. So if there is a strong signal at 380Hz and a weak sized signal at 390Hz, it is sufficient to reproduce the 380Hz signal. The ear won't hear what it is missing.

Most of the major audio compression systems use this masking phenomenon to simplify the process of subband coding. The biggest questions are how to slice up the audible spectrum into subbands and how to identify the strongest signals within the subband. The most sophisticated algorithms do a good job of choosing the right frequencies and representing them with the least number of bits.

12.3 Speech Compression

The different speech compression systems rely on a variety of different mechanisms. Some systems mix and match them in different ways.

The basic speech is first converted into digital sound by sampling the intensity of the sound (also known as the level of *excitation*) at regular intervals. Usually, intervals of about 8,000 to 16,000 times per second suffice. The intensity at these intervals can take a wide range of values, and it is often problematic to simply use a linear quantization system to represent the values. Twelve to sixteen bits of precision are necessary. This produces data at about 96 to 256 kbs.

Compressing this digital stream can be accomplished by strict frequency-based analysis. A Fast Fourier Transform is used to extract the frequencies for a given window or block of time and the significant ones are quantized according to a basic quantization scheme. This is basic subband coding.

Some other systems add more sophisticated analysis of the data stream. One of the most common approaches is to distinguish between *voiced sounds*, *unvoiced sounds*, and *transition sounds*. The voiced sounds are generally pure of pitch and indicated by fairly high intensities or excitation levels and a regular waveform. Many of the vowels are pure voiced sounds ("AAY","Oooh").The unvoiced sounds are produced by the resonance of the vocal cavity. They generally have a lower intensity and a random waveform. They emerge between the voiced sounds. The transition sounds are mixtures that exist as the voiced sounds start up and finish off. Some of the lower-rate coders try to distinguish between these two regions and use different systems to code each of them.

12.4 MPEG and MP3

The MPEG compression standard is aimed at compressing moving pictures, but it includes a fairly sophisticated audio compression algorithm. The system works so well that many people have been using the audio compression system alone without any moving picture. In fact, it is one of the dominant formats on the

Internet right now, known by a variety of different terms like *MP3*, *MPEG Layer 3*, and *Layer 3*.

The MPEG-1 standard itself incorporates three different compression schemes known as layer 1, layer 2, or layer 3. Each layer is slightly more sophisticated than the other and offers slightly better compression. The name *layer* is not particularly appropriate, because each movie would use one of the three systems. They weren't combined as layers. On the other hand, the term has some applicability, because the three approaches are structurally the same. Layer 3 is just a more sophisticated version of Layer 2, and Layer 2 a more sophisticated version of Layer 1.

MPEG-1, Layer 3 audio encoding is also known by the ISO number of the standard: 11172-3.

The different layers provide different amounts of compression. All begin with a pulse coded signal with samples taken at either 33k, 44.1k, or 48k per second. Each layer can produce varying amounts of compression that can be as low as 32k bits per second and as high as 192kbs. The target rate of compression is specified in the header of each file, and it is basically determined by controlling the quantization of the individual frequencies. Higher compression requires more granular quantization, which can reduce the "detail" in the bitstream removing highlights, echoes, and much of the timbre that gives live music its complexity.

The MPEG-1 Layer 3 system is currently the most popular format on the Internet because it does a good job compressing music into small files that tend to run about 64kbs. The standard was developed after many different hearing tests that placed humans in a room and asked them to compare different recordings. This information was used to tune the algorithms to eliminate glitches that allowed the listening audience to identify the worst segments.

The MPEG audio routines are as loosely specified as the MPEG video routines. The bitstream for the data is carefully defined, but there are only rough guidelines on how to do the actual compression required to generate the data in that format. That leaves an opportunity for compression as different laboratories try to come up with the best way of generating the final bitstream. There are many different encoders on the market, and many people have already conducted listening tests to determine which one may be best for each job. It is entirely possible that one encoder may do a better job for orchestral music while another may be superior for rock and roll. Many of these comparisons are subjective, so taste also enters the picture.

Here are the basic steps to the MPEG-1 audio compression functions. The actual detail for each layer will be described afterwards.

Subband Splitting The input signal is split into 32 equal subbands. These are evenly distributed across the audio spectrum, which makes the process relatively easy to implement but less than optimal, because better resolution is often more important in the lower frequencies. Each of these subbands is processed separately to determine the strengths of the signals.

The splitting process is done with an FFT-like matrix-based filter mechanism that slides over a window of the last 512 samples. At each step, another 32 samples are added to the window, which is then weighted by a predefined vector before the frequencies are extracted. This windowing helps reduce strange echoes that can occur during sudden changes in the sound caused by instruments such as cymbals.

Psychoacoustic Modeling The compression algorithm must determine which frequencies are more important than others, and it uses a "psychoacoustic model" to determine which can be heard and which can't. The standard includes two different models of different levels of sophistication.

Determination of Scale Factors The strengths of the frequencies in each subband are normalized by determining a scale factor for the coefficients. That is, it looks for the largest coefficient and then finds an entry in a table of preset scale factors. This effectively removes weaker frequencies, because it reduces them to close to zero after the normalization by the scale factor. Quantization will finish the job.

Conversion to Bits There are still many coefficients that have been scaled. These must be represented as bits. The algorithm allows anywhere from 0 to 15 bits per coefficient, and this controls the resolution. Ideally, the most important coefficients will be represented by more bits so the signal sounds correct. Ten bits allow 1023 different unique levels, which translates into a range of 61.96 decibels. Two bits only allow three different levels and a range of 7 decibels.

To accomplish this, the algorithm starts by allocating only a few bits and then determining which subbands and coefficients need more. The algorithm determines need by calculating a signal-to-noise ratio of the entire block of music. More bits are allocated to the louder, more significant frequencies.

The different layers allocate bits in different ways. The basic Layer 1 uses a fixed number of bits for each block, and this is determined by the user, who chooses a bit rate. Layer 3 allows the different blocks to use more or less than a fixed amount for each block by using a bit reservoir. During a quiet or simple section of the file, fewer bits can be used per block, and the results can be stored up to add more detail later. This can be quite good for speech, which alternates between great detail and silence.

Quantization and Final Packing Once the bits are allocated, each coefficient is scaled by a uniform quantization factor. The results are stored in blocks. In Layer 3, a Huffmanlike code is used to pack the coefficients so the most common ones take up even less space. The extra is put into the bit reservoir, which can be used at appropriate times.

12.5 Conclusion

Audio compression promises to be one of the most important areas of research in the early part of the 21st century because it will form the basis for how major industries blend with the Internet.

The basic techniques, however, are pretty well understood. First, find the significant frequencies using a FFT or other filter bank. Second, decide which frequencies to keep. Voice-only channels can scrimp on high and low frequencies, occasionally mucking up the voice of someone who falls outside of the normal range. Music channels must weigh all candidates and be more judicious. Once the frequencies are chosen, they must be quantized and coded with the appropriate bit stream.

Each of these techniques is well known on its own. The real art is mixing the systems and determining just how and when to cut corners. This will be the basis of much experimentation and listening in the future.

Chapter 13

Fractal Compression

Fractal compression is one of the most mathematical and complicated mechanisms available for compressing data, and as a result it is also one of the most misunderstood. The systems can generate phenomenal compression ratios for images, but only after a significant amount of computation and various trials. The complexity has led some people to joke that the system might as well be called "graduate student compression" because it involves locking a graduate student in a room until he or she happens to find the right-sized set of equations that also reproduces the image.

This joke can be interpreted in many different ways. People with the job of shipping small image files now will probably recoil, because the complexity of the system is too much for them to use regularly. More open-minded individuals, however, will see it as an opportunity to revise and extend the method. The system is not as well understood as the other mechanisms, so there are many improvements and refinements that might be made.

This book can't do a good job summarizing all of the details of the method. The books by Michael Barnsley [Bar93, Bar88, BD86] are much more rigorous and complete, and any serious student of the domain should turn to them.[1] Barnsley is widely regarded as the pioneer of this area of compression, and he has also served as the leader of the firm Iterated Systems, which is commercializing much of the work. The website `www.iterated.com` summarizes the company's work and also offers several excellent white papers about the technology.

At the highest level, fractal compression isn't much different from the other

[1] [BD86] is coauthored with Stephen Demko.

systems described in this book. If y is the data file, then the process of compression is the process of looking for an x such that $f(x) = y$.

Barnsley noted that fractal equations generated incredibly rich scenes from very simple equations. Some of the more famous are the Mandelbrot set and the Julia set. That is, very rich scenes were "compressed" into very simple equations that took up very little data to represent. If these ratios could be duplicated at will, then very good image compression might be achieved. Of course, the problem is that duplicating them can be a challenge. The Mandelbrot and Julia sets began with an equation that turned out to create a very complex and pretty picture. Compressing an arbitrary picture reverses this process and requires finding some equation that can generate it. This isn't always an easy process.

See the work of Benoit Mandelbrot [Man77], who built many of the mathematical foundations of fractal geometry.

There are a wide variety of approaches to fractal geometry, but Barnsley concentrated on one particular one. He looked at iterated affine transformations. That is:

$$\begin{pmatrix} x' \\ y' \end{pmatrix} = \begin{pmatrix} a & b \\ c & d \end{pmatrix} \begin{pmatrix} x \\ y \end{pmatrix} + \begin{pmatrix} e \\ f \end{pmatrix},$$

Ken Falconer offers a look at the mathematical foundations of fractal geometry in his book[Fal90].

where $a, b, c, d, e,$ and f are real numbers that define the affine transformation. The system is called *iterated* because it generates a sequence of (x, y) pairs where $(x_i, y_i) = f(x_{i-1}, y_{i-1})$. If the values of $a, b, c, d, e,$ and f are chosen correctly with the right starting value, then the result can produce a fairly interesting pattern. The sequence of points could define all of the black or white points in an image. The compression ratio can be fairly large, because the function can be defined with only eight values. Clearly this same result can be extended to other dimensions with no trouble.

The biggest challenge is decomposing an image into a number of iterated transformations. Finding this inverse is not an easy process. Much of the work in the area has been devoted to predicting how these iterated transformations behave.[2] Most limit themselves to the linear, affine transformation defined above and search for places where the iterated functions behave in a predictable way. This is often a bit complicated because the functions are also fairly chaotic. Some try to use a codebook that acts like a hash table or an index to the various settings for $a - f$[Ham97, HS93, Kom95, LØ95, Sig96]. Others try to interpolate between different examples by using linear or greater approximation[BMK93, BMK95, Bog94a, Hol91, MH91, Sig97].

The most famous system, however, is the Barnsley Collage Theorem, which describes how to find six points in an image that define self-similarity and use them

[2]These papers are just some of the research looking for simpler ways to solving the inverse fractal transform. [AT89, ATD92, CFMV92, FV94a, FV94b, Har94, HM90, HCF97]

to solve for a, b, c, d, e, and f. Figure 13.1 shows the Black Spleenwort along with two triangles. The fern is constructed with four *strange attractors*, that is, four iterated affine maps. Each dot shows one point along the iteration. It is repeated 60,000 times.

In this case, there are four different affine maps that are chosen randomly. That is, at each step one of the four maps is chosen at random. The values are:

a	b	c	d	d	f	ρ
0	0	0	.16	0	0	.01
.85	.04	$-.04$	.85	0	1.6	.85
.20	$-.26$	.23	.2	0	1.6	.07
$-.15$	.28	.26	.24	0	.44	.07

The value of ρ is the probability that each map will be chosen. These maps were derived by picking self-similar triangles like the two shown in Figure 13.1. If they are chosen correctly, then the points in the strange attractor will bounce around inside the two triangles. This "bouncing around" will be self-similar at many different layers, and this is what produces the neat shape of the fern.

The fractal systems use many the same recursive decomposition used in wavelet methods described in Chapter 9. The image is broken into smaller and smaller blocks until an acceptable fractal equation can be found to represent it. Some simply break the image up into blocks. Others try to match larger blocks with iterated functions before recursively subdividing them. The ideas that work well in wavelet coding often work well in this environment as well [Bog94b, Bon96, CZ97, DACB96, DBC93, DC94, DSC95, RC94, WHK93].

The fractal system also composes multiple transformations to cover the same block. There are ways to combine the results of several transformations so that both add up to look like the original image.

13.1 Conclusion

Fractal compression is still a strange corner of the compression world. The theory is quite elegant and supported by some rigorous mathematical explorations of how functions behave when they are iterated again and again and again. The fact that the system works at all is a bit of a surprise. When it compresses a scene by a huge ratio, it is downright amazing.

In practice, fractal compression is not as impressive. The ratios achieved are similar to better-understood approaches such as the JPEG algorithms. Some make

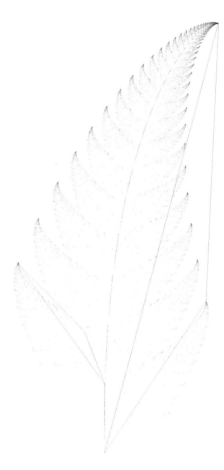

Figure 13.1: This is the Black Spleenwort fern, which is one of the more famous images in fractal geometry. It is generated by four strange attractors calculated with the collage theorem using two self-similar triangles.

the claim that fractal compression offers much deeper resolution than JPEG and offer demonstrations that seem to illustrate it. That is, when the picture is viewed at a large scale and low resolution, then the affine functions are not iterated many times. If a user demands more resolution, then the zoom is accomplished by iterating the function some more. *Et voila*, more points and "detail" appears. The zooming can go deeper and deeper than the original picture.

The problem is that this "detail" isn't the same detail that the original picture being compressed had. It's just simulated detail by the fractal artwork. This can be an advantage in some artistic cases where the user wants something that looks good. The fractal mathematics does a good job of generating something that approximates real life. The new detail, however, is entirely manufactured. Philosophically, this isn't "compressing" an image, it's creating a new one.

The area still is ripe for more research. Doing a good job finding a good way to automatically identify self-similar affine functions for iteration is still an open question. Part of the problem is that many things in life aren't particularily self-similar. While the fern in Figure 13.1 is great, most things in life don't have that repeated symmetry. This makes it harder to come up with the right maps for an image.

Chapter 14

Steganography

Steganography is the science of hiding data. It began with miniature negatives, secret compartments, and disappearing ink and now encompasses a number of mathematical algorithms for hiding data by disguising its true form. Many solutions exist for hiding data as digital photographs, sound files, or text files, and the area is an active branch of research.

Compression is a natural analog to steganography. In the most abstract sense, a compression algorithm detects patterns in data and converts these patterns into nearly random, white noise. Steganography relies on functions that take data and make them assume the patterns and appearance of something else.

That is, if $f_A(x)$ is a compression function that reduces file A to white noise, then $f_A^{-1}(x)$ is a file that converts white noise into something approximating file A. Furthermore, $f_B^{-1}(f_A(x))$ will make file A assume some of the patterns and appearance of file B.

This view works well in theory, but it fails in many cases in practice. Some of the algorithms in this book adapt well to steganography, but many do not. The biggest problem is handling the overhead. Some algorithms do well if the statistical model of the data is separated from the compressed data. Others do not. The ones that allow separation are the ones that do well because this information can be shipped separately to the recipient. The ones that embedded the overhead in the compressed file are not good choices for steganography. In essence, this statistical profile of the file acts as a key for an encryption system and it can't be included or computed on the fly.

The rest of this chapter will discuss several approaches to using compression functions to create files that imitate other files.

14.1 Statistical Coding

Running Huffman compression functions (Chapter 2) or arithmetic compression functions (Chapter 4) in reverse is an ideal way of generating text files that look like someone else's writing. Most English text (and probably the text of other languages as well) has a rich statistical profile that is often highly correlated with the writer and the topic. If f is a statistical compression function, then $f_B^{-1}(f_A(x))$ is an ideal way to generate faux text.

Hugh Kenner and Joseph O'Rourke studied how statistics could generate fake text. They called the result a Travesty.

This approach works better and better with high-order statistics. The use of certain letters varies greatly in English depending on the position in the words. First-order statistics just captures the frequencies of certain letters occurring. Second-order statistics pick up the fact that a 'u' always follows a 'q' and an 'h' often follows a 't'. Real words often appear after third-order statistics are used. Fourth- and fifth-order statistics are quite good, but they often suffer from the lack of source text. In shorter files, many four-letter or five-letter combinations occur once.

The biggest problem with using these statistical functions to hide data is assembling the statistical model for the final data. This source file, A, acts as a key to the process. The most practical solution is for both sender and receiver to agree on some large body of text to act as the model. There are many collections of text on the Net, and it isn't hard to find one.

One form of cryptography would encode words by their page number and position in a particular book. This was known, not surprisingly, as a book cipher.

Here are some samples:

First Order `islhne[hry saeeooisnre uo ' w nala al coehhs pebl e to agboean ce ed cshcenapch nt sibPah ea m n [tmsteoia lahid egnndl y et r yf arleo awe l eo rttnntnnhtohwiseoa a dri 6oc7teit2t lenefe clktoi l mlte r ces. woeiL , misetemd2np eap haled &oolrcc yttr tr,oh en mi elarlbeo tyNunt . syf es2 nrrpmdo,0 reet dadwn'dysg te.ewn1ca-ht eitxrni ntoos xt eCc oh sao vhsoh0mhgr`

Second Order `Thy etheren' ante esthe ales.  icone thers the ase omsictorm s iom.  wactere cut le ce s mo be t Me. Y whes ine odofuion os thore cctherg om tt s d Thm & tthamben tin'ssthe, co westitit odecra fugon tucod. liny Eangem o wen il ea bionBulivethe ton othanstoct itaple`

Third Order `ith eas a tan't genstructin ing butionsmage ruct` `secate expachat thap-res ''Miamproxis is of is a to af st. This there is monst cone usectuabloodes it alu-engettecte por be the andtaly com Bevers gor the Huff-eress. M B G achasion the coduch occomprence mon Quited the ch like bitheres. The`

Fourth Order `captionary. Image and to compression lest con-stance tree. Family for into be mode of bytes in al-gorith a file of that cosition algorithm that word even that a size summarge factal size are:`

`ite position scien Raps.`

`The is are up much length ence, the if the a refsec-ent sec-ent of fits to the crans usuall numberse com-pression`

`A good ways that in algoright. The brase two wants to hidea of English Cash the are compres then matimes formatimes from the data finding pairst. This only be ression o`

The information hidden in this manner is often pretty easy for humans to detect. It may be more difficult for machines. Many computer programs designed to differentiate among random encrypted data, images, and text use statistical models to make the call. They would be fooled by this approach, because the data will assume the statistical profile dictated by a model.

One thousand words take up about 6,000 bytes – much less than the standard image. What's the true equation for the digital age?

14.2 JPEG and JSteg

Images are an easy cover for large amounts of data. It is easy to tweak the RGB coefficients of the different pixels by a small amount and not degrade the image. The coefficients often range from zero to 256, so changing the least-significant bit of a pixel's red coefficient probably won't change the overall value by more than 1%. The lack of effect is helped along by the fact that the pixels are already digital approximations of more detail. The eye is already forgiving the minor blurring introduced when the images were created. What's a bit more?

This channel is a fairly large mechanism for shipping information. It is possible to replace the least-significant bit at every pixel and use one-eighth of the picture's bandwidth for other purposes. Images are also much larger than words, so they have plenty of extra space. In many cases, programmers arrange to use some pseudorandom function to choose the bits that will be tweaked. This spreads around the impact.

This solution for hiding information in images doesn't take advantage of any of the compression algorithms described in the book, so it probably doesn't belong here. To make matters worse, many of the image compression algorithms in this book do not reconstruct the images exactly. This means that they can destroy the information in the least significant bit by accident. The JPEG algorithm, for instance, will replace many of the pixels with approximations. This can ruin any opportunity for hiding information.

The JPEG algorithm, however, can be tweaked on its own to hide information. The solution is to tweak the quantized coefficients of the discrete cosine transform. These can be adjusted by small amounts without changing the image drastically. There will still be some degradation, but it will be minor.

Derek Upham developed this algorithm and named it JSteg. It is widely available on the Net in many of the steganography archives. It will convert a standard targa image into a compressed JPEG file and also include a secret message. The compressed file will be viewable by any JPEG decoder, but only his tool will extract the secret message.

14.3 Quantization

Quantization is an easy compression scheme that can be exploited to hide information. The process is already producing an approximation to the true value so it is often no big deal if an extra bit or two is hidden in with the quantized value. It's all an approximation so the change isn't too obvious.

Image quantization is a popular solution for hiding information. GIF files and other images flood the Net, and there's no reason why extra information can't be inserted into many of these files.

The simplest approach is to use a quantization algorithm to reduce the image data to 2^{n-1} levels that can be indicated with $n-1$ bits. Then, represent the final data with n bits, and hide the extra information as the extra bit in each value of the data. This approach can be fairly easy to detect if the quantized files come with some table used for converting the quantized level values into real data. The GIF files, for instance, have a *pallet* that is just a table with 2^n entries containing the RGB values for each of them. The biggest problem with this solution is that the table will be filled with identical pairs. This is fairly suspicious.

14.4 Grammars

The grammar-based compression algorithms in Chapter 6 can serve as the basis for more sophisticated steganography. One of the more sophisticated approaches for steganography uses grammars to produce files with better verisimilitude [Way95, Way96, Way97]. The grammars create a more sophisticated model of the underlying source, and so they generate a more sophisticated imitation. This extra complexity can be rewarding. In some models, attackers may have just as much trouble as they do attacking strong encryption mechanisms like RSA.

The process of hiding information with a grammar is straightforward. Whenever a decision needs to be made as to which production to choose, use the bits to be hidden to make the call. Here's a sample grammar:

Start	→	**nounverb**
noun	→	Hank ‖ John
verb	→	went hunting **where** ‖ went to the theatre **where**
where	→	in **direction** Canada. ‖ in **direction** California.
direction	→	northern ‖ southern

There are 16 strings in the language defined by this grammar, and this means that four bits can be "hidden" by choosing a string from the grammar. Here's a table that illustrate how the bits '1001' are converted into a string from the grammar.

Bit Hidden	String
none	**Start**
none	**noun verb**
1	John **verb**
0	John went hunting **where**
0	John went hunting in **direction** Canada.
1	John went hunting in southern Canada.

In this example, a zero was hidden by choosing the first production in the grammar for a particular variable, and a one was hidden by choosing the other one. Nothing was hidden when **Start** was replaced by **noun verb**, because there was only one choice. This example had only two choices for each variable, but other grammars may have more. In these cases, more bits can be hidden when more choices are available.

The variables are processed in left-to-right order. This guarantees that the information can be recovered from the sentence. This can be accomplished by a process known as *parsing*, which takes a grammar and determines the sequence of productions that created it. This can be quite a complex process with some grammars, so it makes sense to structure the grammars. Grammars in *Greibach Normal Form*, for instance, have all terminals on the left-hand side of the productions. This example doesn't fall into this case, because the production "**where** $\rightarrow$ in **direction** Canada." has the terminal "Canada" on the right. Grammars in Greibach Normal Form are easier to parse.

Naturally, the quality of the deception depends on the quality of the grammar. What follows is part of the text used to hide the message "I am the walrus." It was produced by a handwritten grammar designed to ape the voiceover from a baseball game. The strikes, balls, and outs are accurate:

```
It's time for another game between the Whappers
and the Blogs in scenic downtown Blovonia .  I've
just got to say that the Blog fans have come to
support their team and rant and rave .  Play Ball
!  Time for another inning .  The Whappers will
be leading off .  Nothing's happened yet to the
Whappers .  Here we go.  Albert Ancien-Regime swings
the baseball bat to stretch and enters the batter's
box .  The pitchers is winding up to throw.  Fans
the air !  He's winding up .  What a blaster .  No
good.  No strike this time .  Here comes the pitch
It's a blaster .  He knocks a line-drive into the
glove of Lefty Clemson The Whappers have two outs
to spare.  The pitcher spits.  Albert Ancien-Regime
swings the baseball bat to stretch and enters the
batter's box .  Love that Baseball game.  Checks
first base .  Nothing.  Winds up and pitches a knuckler
.  OOOh, that's almost in the dirt .  Definitely
a ball .  Checks first base .  Nothing.  Winds up
and pitches a rattler .  High and too inside.  No
strike this time
```

The grammars created by the automatic grammar-based compression programs can be adapted for steganographic purposes. The best solution is to place them into Greibach Normal Form and ship them separately. Another solution is to modify the parsing algorithm to recover the data. Both of these solutions, alas, require some work.

14.5 Conclusions

Steganography is an interesting inverse of compression research and one that may have more importance as the Internet grows more important. Many copyright holders are experimenting with using steganographic solutions to mark documents with hidden messages. These might be simple ownership claims or more complicated unique markers that can be used to track down pirated copies.

Appendix A

Patents

Software patents are controversial, and some of the most controversial are data compression patents. The techniques are very useful to all forms of computer science, so everyone wants to implement them. When people do, they're often shocked to discover that someone else thought of them first and wants royalties for the idea. The problem has been exaggerated by some free-floating collections of code that implement the ideas without mentioning the patents.

This appendix contains the names, numbers, and abstracts of some of the more relevent U.S. patents on compression technology. The list here shouldn't be taken as complete, and no one should use it as the final reference for their search of patent space. Still, the list has many simple advantages, because the patent system is a good repository for ideas. Anyone implementing compression algorithms should check through the list for implementation details.

Also, the reader should be careful not to jump to conclusions. Many patents here focus on only a narrow set of optimizations to the popular algorithms. Many of the patents in the dictionary section, for instance, present different ways to optimize the basic ideas of Lempel and Ziv. In many cases, the patents can be worked around by using slightly different data structures that conveniently circumvent the claims drawn by the inventors.

Programmers should also look for what is not covered in this section. Many software developers don't like software patents and deliberately float their ideas in the public space so they can't be patented. The gzip algorithm, short for Gnu Zip, is a compression program that is freely available and covered by the GNU public license. It was deliberately developed to circumvent patents and provide the world with a patent-free approach that can be used without trouble.

A.1 Statistical Patents

3914586 **Data compression method and apparatus**

Duane E. McIntosh

Issue Date: Oct. 21, 1975

Data compression apparatus is disclosed which is operable in either a bit pair coding mode or a word coding mode depending on the degree of redundancy of the data to be encoded. Consecutive words of data to be encoded are compared and if the words are not identical within a predefined tolerance the data is encoded on a bit pair basis. The bit pair encoding produces transitions in the coded output signals at the beginning of the first of two bit cells which contain a discrete pair of 1's and at the middle of the first of two bit cells which contain a discrete pair of 0's. If the data to be encoded contains a sufficient number of consecutive identical words to permit a greater data compression on a word basis rather than a bit pair basis the first word in the consecutive identical words is encoded on a bit pair basis to identify the bit pattern and a unique transitional pattern incapable of being generated during bit pair coding is generated to identify the number of succeeding words which are identical with the first word.

4494108 **Adaptive source modeling for data file compression within bounded memory**

Glen G. Langdon, Jr. and Jorma J. Rissanen

Issue Date: Jan. 15, 1985

A two-stage single pass adaptive modeling method and means for a finite alphabet first order MARKOV symbol source where the model is used to control an encoder on a per symbol basis thereby enabling efficient compression within a fixed preselected implementation complexity.

4560976 **Data compression**

Steven G. Finn

Issue Date: Dec. 24, 1985

A stream of source characters, which occur with varying relative frequencies, is encoded into a compressed stream of codewords, each having one, two or three subwords, by ranking the source characters by their current frequency of appearance, encoding the source characters having ranks no higher than a first number as one sub-word codewords, source characters having ranks higher than the first number but no higher than a second number as two sub-word codewords, and the remaining source characters as three sub-word codewords. The first number is changed and the second number is recalculated as required by the changing frequencies of the source characters to minimize the length of the stream of codewords.

4612532 **Data compression apparatus and method**
Francis L. Bacon and Donald J. Houde
Issue Date: Sept. 16, 1986

A preferred embodiment of the invention provides a system for the dynamic encoding of a stream of characters. In a preferred embodiment, the system includes an input for receiving a stream of characters, and an output for providing encoded data. The system also includes a provision for storing, accessing, and updating a table (what I call a "followset" table) for each of a plurality of characters, listing candidates for the character which may follow, in the stream, the character with which the table is associated. The system also includes a provision for furnishing at the output, for a given character in the stream at the input, a signal indicative of the position, occupied by the given character, in the followset character which immediately precedes the given character in the stream of the input. A preferred embodiment of the invention also provides a system for decoding the encoded data, wherein the decoder utilizes a followset table.

4853696 **Code converter for data compression/decompression**
Amar Mukherjee
Issue Date: Aug. 1, 1989

A code converter has a network of logic circuits connected in reverse binary tree fashion with logic paths between leaf nodes and a common root node. Characters are compressed from standard codes to variable-length Huffman code by pulse applying connections to the paths from a decoder. An OR-gate is connected to "1" branches of the network to deliver the output code. Characters are decompressed from Huffman to standard codes by connection of the Huffman code to control the clocked logic circuits to deliver a pulse from the root node to one of the inputs of an encoder. A feedback loop connects the root node and the path end nodes to initiate the next conversion. Alternative embodiments have decoder staging to minimize delay and parallel compressed code output.

5023611 **Entropy encoder/decoder including a context extractor**

Christosdoulos Chamzas and Donald L. Duttweiler
Issue Date: June 11, 1991

In entropy, e.g., arithmetic or adaptive Huffman, encoding/decoding a context based on prior symbols is needed to provide accurate predictions of symbols to be encoded/decoded. Encoding/decoding efficiency is improved by employing an adaptive context extractor. The adaptive context extractor automatically adjusts the configuration of the lag intervals used to define the context. This is realized by adaptively incorporating into the context configuration at least one lag interval found to have a "good" predictive value relative to the particular symbol stream being encoded/decoded. The context configuration adjustment is such that the at least one found lag interval is exchanged with the lag interval currently in an at least one so-called floating predictor position.

5025258 **Adaptive probability estimator for entropy encoding/decoding**
Donald L. Duttweiler
Issue Date: June 18, 1991

In entropy, e.g. arithmetic, encoding and decoding, probability estimates are needed of symbols to be encoded and subsequently decoded. More accurate probability estimates are obtained by controllably adjusting the adaptation rate of an adaptive probability estimator. The adaptation rate is optimized by matching it to the actual probability values being estimated. In particular, the adaptation rate is optimized to be proportional to the inverse of the smallest value probability being estimated. Consequently, if the probability values being estimated are not small a "fast" adaption rate is realized and if the probability values being estimated are small a necessarily slower adaptation rate is realized.

5404140 **Coding system**
Fumitaka Ono, Tomohiro Kimura, Masayuki Yoshida, and Shigenori Kino
Issue Date: April 4, 1995

A coding system comprises the comparing circuit which compares a magnitude of the range on the number line which is allocated to the most probability symbol with a magnitude of the fixed range on the number line which is allocated to the Less Probability Symbol. If the range allocated to the MPS is smaller than that to the LPS, and when the symbol is the MPS, the range allocated to the LPS is generated. If the range allocated to the MPS is smaller than that to the LPS, and when the symbol is the LPS, the range allocated to the MPS is generated. By the system, a coding efficiency is improved especially when a probability of occurrence of LPS (Less Probability Symbol) is approximate to 1/2.

A.2 Dictionary Algorithm Patents

The dictionary algorithm patents have been one of the most contentious areas of software patent law, in part because they've been widely implemented, often by people who didn't know about some of the patents. The most successful ones may be the ones assigned to STAC (4,701,745; 5,016,009; etc.), which were successfully defended in a court case that cost Microsoft millions of dollars in royalties. Even the original patents by Lempel and Ziv have been quite controversial because the same algorithm forms the basis of the almost ubiquitous GIF file format. UNISYS, which controls the patents, has stated that they have been interested only in approaching commercial software developers for a license to the patent. While this doesn't affect the average folks using a browser, it is of little consolation to the software developers trying to use the GIF format.

4021782 **Data compaction system and apparatus**
John S. Hoerning
Issue Date: May 3, 1977

A data compaction system and apparatus is disclosed which, in the preferred embodiment, includes a high speed compaction controller utilizing both read only storage and read-write storage. A compaction device according to the present invention could then be placed upon both ends of a transmission line, the data received by a compaction unit at one end of the line from whatever apparatus wished to transmit data on the line, the data compacted within the compaction unit according to the present invention, the data transmitted on the line to a compaction unit on the other end of the line, the data decompacted, and the data provided to whatever apparatus wished to receive the data. Data received by the compaction device according to the present invention in a fixed length, fixed number base, coded manner would then be compacted by altering the expression of the data, as by altering the number bases in which the data is expressed and by switching between number bases. Thus, it has been found that expressing the data as a string of characters of varying lengths and varying number bases, that is characters not all expressed in the same number base, shortens the overall length of the data transmitted. This is true even if the length of certain characters may be increased by the techniques according to the present invention. Also, prior character and prior record comparisons according to the present invention significantly enhance the compaction ability of the present invention.

4054951 **Data expansion apparatus**

Rory D. Jackson and Willi K. Rackl

Issue Date: Oct. 18, 1977

When a data stream includes long sections of data that are repeated periodically, storage space may be saved by not including full repetitions of such sections in the storage. However, when the data is to be read from storage for utilization, the omitted repetitious sections must be inserted. This is accomplished by providing hardware apparatus which recognizes a particular flag occurring in the stored data. After recognizing the flag, the expansion apparatus interprets the next piece of information in the data stream as being the storage address of the start of a section of data that is to be inserted into the data stream; the next piece of information is interpreted as being the length of the section of data to be inserted; and the next following piece of information is the number of times that the section of data is to be inserted. The apparatus will respond to the flag and its associated indicators by inserting the appropriate data section the indicated number of times.

4412306 **System for minimizing space requirements for storage and transmission of digital signals**

Edward W. Moll

Issue Date: Oct. 25, 1983

Digital information is encoded prior to storage for minimizing the space required for storage of the digital information. The data is compared with prior received data such as to detect repetition. When repetition is detected, a code indication of the period of repetition and the duration of repetition is inserted in the stored data. Information is restored utilizing decoding networks responsive to the inserted codes which cause reinsertion of the repetitions.

4464650 **Apparatus and method for compressing data signals and restoring the compressed data signals**

Willard L. Eastman, Abraham Lempel, Jacob Ziv, and Martin Cohn

Issue Date: Aug. 7, 1984

A compressor parses the input data stream into segments where each segment comprises a prefix and the next symbol in the data stream following the prefix.

The prefix of a segment is the longest match with a previously parsed segment of the data stream. The compressor constructs a search tree data base to effect the parsing and to generate a pointer for each segment pointing to the previous segment matching the prefix. The search tree comprises internal nodes including a root and external nodes denoted as leaves. The nodes are interconnected by branches representative of symbols of the alphabet. Each parsed segment of the input data is represented by a path from the root to a leaf. The tree is adaptively constructed from the input data such that as each new segment is parsed, one new internal node of the tree is created from a leaf and new leaves are defined, one for each symbol already encountered by the encoder plus an additional branch to represent all potential but unseen symbols. The compressor transmits a leaf pointer signal for each parsed segment representative of the prefix thereof and the suffixed symbol of the alphabet. A decompressor constructs an identical search tree in response to the received leaf pointers so as to reconstitute the original data stream.

4558302 **High speed data compression and decompression apparatus and method**

Terry A. Welch

Issue Date: Dec. 10, 1985

A data compressor compresses an input stream of data character signals by storing in a string table strings of data character signals encountered in the input stream. The compressor searches the input stream to determine the longest match to a stored string. Each stored string comprises a prefix string and an extension character where the extension character is the last character in the string and the prefix string comprises all but the extension character. Each string has a code signal associated therewith and a string is stored in the string table by, at least implicitly, storing the code signal for the string, the code signal for the string prefix and the extension character. When the longest match between the input data character stream and the stored strings is determined, the code signal for the longest match is transmitted as the compressed code signal for the encountered string of characters and an extension string is stored in the string table. The prefix of the extended string is the longest match and the extension character of the extended string is the next input data character signal following the longest match. Searching through the string table and entering extended strings therein is effected by a limited search hashing procedure. Decompression is effected by a decompressor that receives the compressed code signals and generates a string table similar to that constructed by the compressor to effect lookup of received code signals so as to recover the data character signals comprising a stored string. The decompressor string table is updated by storing a string having a prefix in accordance with a prior received code signal and an extension character in accordance with the first character of the currently recovered string.

4597057 **System for compressed storage of 8-bit ASCII bytes using coded strings of 4 bit nibbles**

Craig A. Snow

Issue Date: June 24, 1986

Standard ASCII coded text is divided into alpha, numeric, and punctuation tokens. Each token is converted to a string of four-bit nibbles. One nibble is coded to identify the type of token. Additional nibbles are coded to identify the location, if any, of a corresponding alpha or punctuation token in a global dictionary. If no corresponding alpha token is in the dictionary, an alpha token is divided into prefixed, suffixes, and a stem. The location of any prefixes in a table of prefixes, suffixes in a table of suffixes, and the number, and location of corresponding individual characters in a table, of the remaining stem are then coded and stored as part of the string of four-bit nibbles for the alpha tokens. Numeric tokens are stored as a string of four-bit nibbles in which the first nibble identifies the type of token, the next nibble the length, followed by a nibble for each of the digits.

4701745 **Data compression system**

John R. Waterworth

Issue Date: Oct. 20, 1987

A data compression system includes an input store (1) for receiving and storing a plurality of bytes of data from an outside source. Data processing means for processing successive bytes of data from the input store includes circuit means (21-25) operable to check whether a sequence of bytes is identical with a sequence of bytes already processed, output means (27) operable to apply to a transfer medium (12) each byte of data not forming part of such an identical sequence, and an encoder (26) responsive to the identification of such a sequence to apply to the transfer means (12) an identification signal which identifies both the location in the input store of the previous occurrence of the sequence of bytes and the number of bytes in the sequence.

4730348 **Adaptive data compression system**

John E. MacCrisken

Issue Date: March 8, 1988

A data compression system for increasing the speed of data transmission system over a communication channel with a predefined data transmission rate. The system has two data compression units – one on each end of the channel, coupled to first and second data processing systems. Input data from either data processing system is encoded using a selected one of a plurality of encoding tables, each of which defines a method of encoding data using codes whose length varies inversely with the frequency of units of data in a predefined set of data. Whenever an analysis of the encoded data indicates that the data is not being efficiently compressed, the system invokes a table changer for selecting from among the encoding tables the one which minimizes the bit length of the encoded data for a preselected sample of the input data. If a new table is selected, a table change code which corresponds to the selected table is added to the encoded data. Also, a dynamic table builder builds a new encoding table to be including in the set of available encoding tables using a preselected portion of the previously encoded input data which an analysis of the encoded data indicates that a new encoding table will enhance compression. Each data compression unit includes a data decoder for decoding encoded data sent over the channel by the other unit. Thus the data decoder uses a set of decoding tables corresponding to the encoding tables, means for selecting a new table when a table change code is received, and means for building a new decoding table when it receives a table change code which indicates that the encoded data following the table change code was encoded using a new encoding table.

4814746 **Data compression method**
Victor S. Miller and Mark N. Wegman
Issue Date: March 21, 1989

Communications between a Host Computing System and a number of remote terminals is enhanced by a data compression method which modifies the data compression method of Lempel and Ziv by addition of new character and new string extensions to improve the compression ratio, and deletion of a least recently used routine to limit the encoding tables to a fixed size to significantly improve data transmission efficiency.

4906991 **Textual substitution data compression with finite length search windows**
Edward R. Fiala and Daniel H. Greene
Issue Date: March 6, 1990

In accordance with the present invention source data is encoded by literal codewords of varying length value, with or without the encoding of copy codewords of varying length and displacement value. Copy codeword encoding is central to textual substitution-style data compression, but the encoding of variable length literals may be employed for other types of data compression.

4955066 **Compressing and decompressing text files**
Leo A. Notenboom
Issue Date: Sept. 4, 1990

A method of compressing a text file in digital form is disclosed. A full text file having characters formed into phrases is provided by an author. The characters are digitally represented by bytes. A first pass compression is sequentially followed by a second pass compression of the text which has previously been compressed. A third or fourth level compression is serially performed on the previously compressed text. For example, in a first pass, the text is run-length compressed. In a second pass, the compressed text is further compressed with key phrase compression. In a third pass, the compressed text is further compressed with Huffman compression. The compressed text is stored in a text file having a Huffman decode tree, a key phrase table, and a topic index. The data is decompressed in a single pass and provided one line at a time as an output. Sequential compressing of the text minimizes the storage space required for the file. Decompressing of the text is performed in a single pass. As a complete line is decompressed, it is output rapidly, providing full text to a user.

4988998 **Data compression system for successively applying at least two data compression methods to an input data stream**
John T. O'Brien
Issue Date: Jan. 29, 1991

The improved data compression system concurrently processes both strings of repeated characters and textual substitution of input character strings. In this system, the performance of data compression techniques based on textual substitution are improved by the use of a compact representation for identifying instances in which a character in the input data stream is repeated. This is accomplished by nesting a run length encoding system in the textual substitution system. This structure adds the recognition of runs of a repeated character before the processor performs the textual substituted data compression operation. A further performance improvement is obtained by expanding the alphabet of symbols stored in the compressor's dictionary to include both the characters of the input data stream and repeat counts which indicate the repetition of a character. The handling of these repeat counts by the textual substitution based compression technique is no different than the handling of characters, or certain modifications are made in the handling of repeat counts.

5001478 **Method of encoding compressed data**
Michael E. Nagy
Issue Date: March 19, 1991

A method of encoding compressed data for transmission or storage. The method includes converting the input data stream into a sequence of literal references, history references, and lexicon references. Each literal reference includes a literal identifier. Each history reference includes a history identifier. Each lexicon reference includes a lexicon identifier. The history references and lexicon references specify the location of strings of symbols in a history buffer and lexicon, respectively. The encoding method of the present invention encodes the location information in an optimal manner.

5016009 **Data compression apparatus and method**
Douglas L. Whiting, Glen A. George, and Glen E. Ivey
Issue Date: May 14, 1991

An apparatus and method for converting an input data character stream into a variable length encoded data stream in a data compression system. The data compression system includes a history array means. The history array means has a plurality of entries and each entry of the history array means is for storing a portion of the input data stream. The method for converting the input data character stream includes the following steps. Performing a search in a history array means for the longest data string which matches the input data string. If the matching data string is found within the history buffer means, the next step includes encoding the longest matching data string found by appending to the encoded data stream a tag indicating the longest matching data string was found and a string substitution code. If the matching data string is not found within the history array means, the next step includes encoding the first character of the input data string by appending to the encoded data stream a raw data tag indicating that no matching data string was found and the first character of the input data string.

5049881 **Apparatus and method for very high data rate-compression incor-**

porating lossless data compression and expansion utilizing a hashing technique
Dean K. Gibson and Mark D. Graybill
Issue Date: Sept. 17, 1991

A method and apparatus for compressing digital data that is represented as a sequence of characters drawn from an alphabet. An input data block is processed into an output data block composed of sections of variable length. Unlike most prior art methods which emphasize the creation of a dictionary comprised of a tree with nodes or a set of strings, the present invention creates its own pointers from the sequence characters previously processed and emphasizes the highest priority on maximizing the data rate-compression factor product. The use of previously input data acting as the dictionary combined with the use of a hashing algorithm to find candidates for string matches and the absence of a traditional string matching table and associated search time allows the compressor to very quickly process the input data block. Therefore, the result is a high data rate-compression factor product achieved due to the absence of any string storage table and matches being tested only against one string.

5051745 **String searcher, and compressor using same**
Phillip W. Katz
Issue Date: Sept. 24, 1991

Methods and apparatus for string searching and data compression. In the string search method and apparatus pointers to the string to be searched are indexed via a hashing function and organized according to the hashing values of the string elements pointed to. The hashing function is also run on the string desired to be found, and the resulting hashing value is used to access the index. If the resulting hashing value is not in the index, it is known that the target string does not appear in the string being searched. Otherwise the index is used to determine the pointers which correspond to the target hashing value, these pointers pointing to likely candidates for matching the target string. The pointers are then used to sequentially compare each of the locations in the string being searched to the target string, to determine whether each location contains a match to the target string. In the method and apparatus for compressing a stream of data symbols, a fixed length search window, comprising a predetermined contiguous portion of the symbol stream, is selected as the string to be searched by the string searcher. If a string to be compressed is found in the symbol stream, a code is output designating the location within the search window of the matching string and the length of the matching string.

5109433 **Compressing and decompressing text files**

Leo A. Notenboom

Issue Date: April 28, 1992

A method of compressing a text file in digital form is disclosed. A full text file having characters formed into phrases is provided by an author. The characters are digitally represented by bytes. A first pass compression is sequentially followed by a second pass compression of the text which has previously been compressed. A third or fourth level compression is serially performed on the previously compressed text. For example, in a first pass, the text is run-length compressed. In a second pass, the compressed text is further compressed with key phrase compression. In a third pass, the compressed text is further compressed with Huffman compression. The compressed text is stored in a text file having a Huffman decode tree, a key phrase table, and a topic index. The data is decompressed in a single pass and provided one line at a time as an output. Sequential compressing of the text minimizes the storage space required for the file. Decompressing of the text is performed in a single pass. As a complete line is decompressed, it is output rapidly, providing full text to a user.

5126739 **Data compression apparatus and method**

Douglas L. Whiting, Glen A. George, and Glen E. Ivey

Issue Date: June 30, 1992

An apparatus and method as disclosed for converting an input data character stream into a variable length encoded data stream in a data compression system. The data compression system includes a history array means. The history array means has a plurality of entries and each entry of the history array means is for storing a portion of the input data stream. The method for converting the input data character stream includes the following steps. Performing a search in a history array means for the longest data string which matches the input data string. If the matching data string is found within the history buffer means, the next step includes encoding the longest matching data string found by appending to the encoded data stream a tag indicating the longest matching data string was found and a string substitution code. If the matching data string is not found within the history array means, the next step includes encoding the first character of the input data string by appending to the encoded data stream a raw data tag indicating that no matching data string was found and the first character of the input data string.

5140321 **Data compression/decompression method and apparatus**

Robert K. Jung
Issue Date: Aug. 18, 1992

A method and apparatus for compressing digital data uses data which has been previously compressed as a dictionary of substrings which may be replaced in an input data stream. The method and apparatus uses a hash table to take advantage of principles of locality and probability to solve the maximal matching substring problem inherent in this type of compressing apparatus, most of the time. The hash table consists of first-in, first-out (FIFO) collision chains of fixed, uniform numbers of pointers to substrings of data already compressed which potentially match an input substring. A companion decompressing method and apparatus receives compressed data from the compressing apparatus and expand that data back to its original form.

5155484 **Fast data compressor with direct lookup table indexing into history buffer**
Lloyd L. Chambers, IV
Issue Date: Oct. 13, 1992

A cooperating data compressor, compressed data format, and data decompressor. The compressor compresses an input data block (HB) to a compressed data block having the format. The decompressor decompresses the compressed data block to restore the original data block. The compressor has a direct lookup table (DLT) of 2^8N entries, each indexable by N bytes at a current HB location and identifying a target HB location. The compressor determines whether a target string at the target HB location matches a current string at the current HB location. If they do not match, the compressor outputs a literal representing a datum at the current location. If they match, the compressor outputs a vector from the current location to the target string. Compression speed is maximized by the direct addressing of the DLT by the current N bytes in the HB. Decompression speed is maximized by state machine operation according to literal/vector indicators, special case string length codes, and special case offset codes in the compressed data format.

5179378 **Method and apparatus for the compression and decompression of data using Lempel-Ziv based techniques**
N. Ranganathan Selwyn Henriques and
Issue Date: Jan. 12, 1993

A method and apparatus for compressing and decompressing text and image data by forming fixed-length codewords from a variable number of data symbols. Data symbols are shifted into registers in the first half of a buffer, while an uncompressed string of data symbols are shifted into registers in the second half of the buffer. A systolic array of processors compares each data symbol in the second half of the buffer with each data symbol in the first half of the buffer. Each processor compares pairs of data symbols, and selectively passes the data symbols to an adjacent processor. A fixed-length output is provided indicating the length and the starting point of the longest substring in the first half of the buffer that matches a substring from the second half of the buffer. The matched data symbols in the second half of the buffer and the data symbol immediately following the matched data symbols are then shifted into the first half of the buffer, and uncompressed data symbols are then shifted into the second half of the buffer. A preselected shift register in the first half of the buffer provides a fixed-length output indicating the symbol that immediately follows the last matched data symbol. The length and the starting point information and the last symbol information are assembled to form a codeword having a predetermined length. The codeword is stored in memory and can be later retrieved and decompressed to provide the original string of data symbols.

5229768 **Adaptive data compression system**

Kasman E. Thomas

Issue Date: July 20, 1993

A system for data compression and decompression is disclosed. A series of fixed length overlapping segments, called hash strings, are formed from an input data sequence. A retrieved character is the next character in the input data sequence after a particular hash string. A hash function relates a particular hash string to a unique address in a look-up table (LUT). An associated character for the particular hash string is stored in the LUT at the address. When a particular hash string is considered, the content of the LUT address associated with the hash string is checked to determine whether the associated character matches the retrieved character following the hash string. If there is a match, a Boolean TRUE is output; if there is no match, a Boolean FALSE along with the retrieved character is output. Furthermore, if there is no match, then the LUT is updated by replacing the associated character in the LUT with the retrieved character. The process continues for each hash string until the entire input data sequence is processed. The method of decompression includes the steps of initializing a decompression LUT to mirror the initial compression LUT and receiving a representational form output from the compressor. The representational form is generally analyzed one character at a time. If the character is a Boolean TRUE, then the content of the LUT addressed by the most recently decoded hash string is output. Otherwise, if the character is a Boolean FALSE, the next character (exception character) in the representational form is output and the content of the LUT addressed by the most recently decoded hash string is output.

A.3 Arithmetic Algorithm Patents

4122440 **Method and means for arithmetic string coding**

Glenn George Langdon, Jr. and Jorma Johannen Rissanen

Issue Date: Oct. 24, 1978

There is disclosed a method and means for compacting (encoding) and decompacting (decoding) binary bit strings which avoids the blocking of string elements required by Huffman coding and the ever increasing memory as is the case in simple enumerative coding. The method and means arithmetically encodes successive terms or symbols in a symbol string $s = a_i a_j \ldots$, in which each new term a_k in a source alphabet of N symbols gives rise to a new encoded string $C(sa_k)$ and a new length indicator $L(sa_k)$. The method and means comprises the steps of forming $L(sa_k)$ from the recursion $L(sa_k) = L(s) + l(a_k)$, where $l(a_k)$ is a rational approximation of $\log_2 1/p(a_k)$, $p(a_k)$ being the a priori probability of occurrence of a_k, and $l(a_k)$ being so constrained that the Kraft inequality is satisfied: *[Figure]* AND FORMING $C(sa_k)$ from the recursion $C(s) + [P_{k-1}\, 2^L(sa.sbsp.k)]$,where: *[Figure]* AND WHERE $P_{k-1}$is the cumulative probability of occurrence of an arbitrary ordering of all symbols. Instead of assigning a'priori code words to each symbol as in Huffman coding, the method and means ascertains the new encoded string $C(sa_k)$ from the ordinal number (position number k) of the symbol a_k, in the arbitrary ordering, the value of the fractional part of $L(sa_k)$, and the last few bits of the previous compressed symbol string $C(s)$. This results in the generation of two quantities i.e. the encoded string of compressed characters and the fractional length indicators. The use of only a few bits of the last encoded string $C(s)$ for determining $C(sa_k)$ limits the requisite memory to a constant value. During each encoding cycle, the least significant number of bits of $C(s)$ are shifted out as determined by the integrer part of $L(sa_k)$.

4286256 **Method and means for arithmetic coding utilizing a reduced number of operations**

Glen G. Langdon, Jr. and Jorma J. Rissanen
Issue Date: Aug. 25, 1981

A method and means of arithmetic coding of conditional binary sources permitting instantaneous decoding and minimizing the number of encoding operations per iteration. A single shift and subtract operation for each encoding cycle can be achieved if an integer valued parameter representative of a probability interval embracing each source symbol relative frequency is used for string encoding and control. If the symbol being encoded is the most probable, then nothing is added to the arithmetic code string. However, an internal variable is updated by replacing it with an augend amount. If the updated internal variable has a leading zero, then both it and the code string are shifted left by one position. If the symbol being encoded is the least probable, then a computed augend is added to the code string and the code string is shifted by an amount equal to the integer valued parameter.

4295125 **Method and means for pipeline decoding of the high to low order pairwise combined digits of a decodable set of relatively shifted finite number of strings**
Glen G. Langdon, Jr.
Issue Date: Oct. 13, 1981

An apparatus for ensuring continuous flow through a pipeline processor as it relates to the serial decoding of FIFO Rissanen/Langdon arithmetic string code of binary sources. The pipeline decoder includes a processor (11, 23) and a finite state machine (21, FSM) in interactive signal relation. The processor generates output binary source signals (18), status signals (WASMPS, 31) and K component/K candidate next integer-valued control parameters (L0, k0; L1, k1; 25). These signals and parameters are generated in response to the concurrent application of one bit from successive arithmetic code bits, a K component present integer-value control parameter (52) and K component vector representation (T, TA) of the present internal state (51) of the associated finite state machine (FSM). The FSM makes a K-way selection from K candidate next internal states and K candidate next control parameters. This selection uses no more than $K^2 + K$ computations. The selected signals are then applied to the processor in a predetermined displaced time relation to the present signals in the processor. As a consequence, this system takes advantage of the multi-state or "memory" capability of an FSM in order to control the inter-symbol influence and facilitate synchronous multi-stage pipeline decoding.

4463342 **Method and means for carry-over control in the high order to low**

order pairwise combining of digits of a decodable set of relatively shifted finite number strings

Glen G. Langdon, Jr. and Jorma J. Rissanen

Issue Date: July 31, 1984

Carry-over control in strings resulting from the high to low order combining of two binary number strings is obtained through the insertion of a control character within the resultant string after detecting a run of consecutive 1's. Upon subsequent accessing and decomposition of the resultant string, the control character causes string decomposition to operate for a number of cycles in a carry correction mode. If the control character indicates that a carry has rippled through the n lesser significant positions of the resultant string, then upon decomposition, those "n" consecutive 1's are changed to 0's, and a 1 is added to the least significant position in the string preceding the control character. If the control character indicates no carry occurrence, then it is merely deleted from the string. The control of carries in this manner permits the generation of arithmetic string compression code sequences in an instantaneous FIFO pattern with only a modest reduction of compression efficiency. Relatedly, the encoder and carry suppressor is shown in FIG. 2[1], while the carry corrector and decoder is shown in FIG. 4.

4467317 **High-speed arithmetic compression coding using concurrent value updating**

Glen G. Langdon, Jr. and Jorma J. Rissanen

Issue Date: Aug. 21, 1984

A method and apparatus for recursively generating an arithmetically compressed binary number stream responsive to the binary string from conditional sources. Throughput is increased by reducing the number of operations required to encode each binary symbol so that only a single shift of k bits is required upon receipt of each least-probable symbol or an "add time", followed by a decision and a one-bit shift in response to each most-probable symbol encoding. The concurrent augmentation of the compressed stream and an internal variable involves only the function of a probability interval estimate of the most-probable symbol, and not upon the past encoding state of either variable $(2^{-k}, 49, 63, C, T)$. Each binary symbol may be recovered by subtracting 2^{-k} from the q-most-significant bits of the compressed stream and testing the leading bit of the difference.

[1] See original patent for figures.

4633490 **Symmetrical optimized adaptive data compression/ transfer/ decompression system**

Gerald Goertzel and Joan L. Mitchell

Issue Date: Dec. 30, 1986

Data compression for transfer (storage or communication) by a continuously adaptive probability decision model, closely approaches the compression entropy limit. Sender and receiver perform symmetrical compression/decompression of binary decision n according to probabilities calculated independently from the transfer sequence of $1 \ldots n-1$ binary decisions. Sender and receiver dynamically adapt the model probabilities, as a cumulative function of previously presented decisions, for optimal compression/decompression. Adaptive models for sender and receiver are symmetrical, to preserve data identity; transfer optimization is the intent. The source model includes a state generator and an adaptive probability generator, which dynamically modify the coding of decisions according to state, probability and bit signals, and adapt for the next decision. The system calculates probability history for all decisions, including the current decision, but uses probability history for decision $n-1$ (the penultimately current decision) for encoding decision n (the dynamically current decision). The system, separately at source and destination, reconfigures the compression/expansion algorithm, up to decision $n-1$, codes each decision in the data stream optimally, according to its own character in relation to the calculated probability history, and dynamically optimizes the current decision according to the transfer optimum of the data stream previously transferred. The receiver operates symmetrically to the sender. Sender and receiver adapt identically, and adapt to the same decision sequence, so that their dynamically reconfigured compression-expansion algorithms remain symmetrical–even though the actual algorithms may change with each decision as a function of dynamic changes in probability history.

4652856 **Multiplication-free multi-alphabet arithmetic code**

Kottappuram M. A. Mohiuddin and Jorma J. Rissanen

Issue Date: March 24, 1987

Method and apparatus which cyclically generate a compressed, arithmetically-coded binary stream in response to binary occurrence counts of symbols in an uncoded string. The symbols in the uncoded string are drawn from a multi-character alphabet which is not necessarily a binary one. Coding operations and hardware are simplified by deriving from the binary occurrence counts an estimate of the probability of each unencoded symbol at its precise lexical location. The probability estimation eliminates any requirement for division or multiplication by employing magnitude-shifting of the binary occurrence counts. The encoded stream is augmented by the estimated symbol probability at the same time that an internal variable is updated with an estimate of the portion of a probability interval remaining after coding the current symbol, the interval estimate being obtained from the left-shifted occurrence counts. Decoding is the dual of encoding. The unencoded symbol stream is extracted, symbol-by-symbol, by subtracting the estimated symbol probability that comes closest to, but does not exceed the magnitude of the compressed stream, re-estimating the symbol probabilities based upon the decoding, and testing the difference of the subtraction against the re-estimated probability.

4792954 **Concurrent detection of errors in arithmetic data compression coding**

Ronald B. Arps and Ehud D. Karnin

Issue Date: Dec. 20, 1988

Method and means for detecting any single errors introduced into an arithmetic data compression code string as a result of coding, transmission, or decoding through testing according to a modulo-n function, n being an odd number not equal to $+1$ or -1, an arithmetic compression coded data stream $C'(s)$ genmerated by n-scaling an arithmetically recursive function that produces a non-scaled arithmetically data compressed code stream $C(s)$ as a number in the semi-open coding range [0,1].

4891643 **Arithmetic coding data compression/de-compression by selectively employed, diverse arithmetic coding encoders and decoders**

Joan L. Mitchell and William B. Pennebaker

Issue Date: Jan. 2, 1990

A data compression/de-compression system includes a first arithmetic coding encoder, characterized by a first set of encoding conventions, which encoder generates a code stream that points to an interval along a number line in response to decision event inputs. The code stream can be adjusted to point to the same interval as code streams generated by one or more other arithmetic coding encoders characterized by encoding conventions differing in some way from those in the first set. In a binary context, optimal hardware encoders increment or decrement the value of the code stream in response to each occurrence of a more probable decision event while optimal software so changes the code stream value for each occurrence of a less likely event. According to the invention, the code streams for optimal hardware encoders and optimal software encoders are made either identical or compatible to enable similar decoding for each. Identical or compatible code streams are obtained from encoders having different event sequence or symbol ordering along intervals on the number line. Moreover, various hardware and software decoders–with respective symbol ordering and other conventions–can be used in conjunction with encoders having respective conventions, wherein each decoder retrieves the same sequence of decisions for a code stream pointing to a given interval. In both encoding and decoding, the present invention overcomes finite precision problems of carry propagation and borrow propagation by handling data in bytes and bit stuffing at byte boundaries and by pre-borrowing as required.

4905297 **Arithmetic coding encoder and decoder system**

Glen G. Langdon, Jr., Joan L. Mitchell, William B. Pennebaker, and Jorma J. Rissanen

Issue Date: Feb. 27, 1990

Apparatus and method for compressing and de-compressing binary decision data by arithmetic coding and decoding wherein the estimated probability Qe of the less probable of the two decision events, or outcomes, adapts as decisions are successively encoded. To facilitate coding computations, an augend value A for the current number line interval is held to approximate one by renormalizing A whenever it becomes less than a prescribed minimum AMIN. When A is renormalized, the value of Qe is up-dated. The renormalization of A and up-dating of Qe are preferably based on a single-bit test. Also, each Qe value is preferably specified as a 12-bit value having the least significant bit set to 1 and having no more than four other bits set to 1. The number of Qe values in the 1/4 to 1/2 probability range is enhanced to improve coding efficiency. A decision coding parameter of preferably six bits indicates the sense of the more probable symbol (MPS) in one bit and identifies a corresponding Qe value with the remaining five bits. In addition to probability adaptation, the present invention discloses an allocation of bits in a code stream register in which preferably two spacer bits are inserted between a next byte portion (which contains a byte of data en route to a buffer) and a fractional portion which may be involved in further computation. With the two spacer bits, any code greater than or equal to Hex ‘CO’ which follows a Hex ‘FF’ byte is illegal for data and therefore provides for an escape from the code stream. The two spacer bits also reduce the number of stuff bits inserted to account for carry or borrow propagation. Encoding and decoding can be performed interchangeably by hardware or software which feature differing coding conventions.

4901363 **System for compressing bi-level data**

Kazuharu Toyokawa

Issue Date: Feb. 13, 1990

A system is disclosed for arithmetically compressing bi-level data of an image by sampling history pels, which are adjacent pels to a current pel to be processed, and pels so separated from the adjacent pels as to adapt to a period of a dither matrix for a dither image. From the sampled pels it is determined whether the sampled image has a dither dominant image pattern or a text-graphic dominant image pattern and an appropriate signal is generated, a first signal representing that the image has the dither dominant image pattern, a second signal representing that the image has the text-graphic dominant image pattern, and a third signal representing that the image is not classified as either of these image patterns. The first, second and third signals are applied to an up-down counter which accumulates the occurrences of the signals to produce, when the accumulated value exceeds an upper threshold or a lower threshold, a control signal for selectively switching access to first and second statistical tables, which store, at each entry, the more probable symbol and the probability of the less probable symbol for the image of the text-graphic image and the dither image, respectively.

4905297 **Arithmetic coding encoder and decoder system**

Glen G. Langdon, Jr., Joan L. Mitchell, William B. Pennebaker, and Jorma J. Rissanen

Issue Date: Feb. 27, 1990

Apparatus and method for compressing and de-compressing binary decision data by arithmetic coding and decoding wherein the estimated probability Qe of the less probable of the two decision events, or outcomes, adapts as decisions are successively encoded. To facilitate coding computations, an augend value A for the current number line interval is held to approximate one by renormalizing A whenever it becomes less than a prescribed minimum AMIN. When A is renormalized, the value of Qe is up-dated. The renormalization of A and up-dating of Qe are preferably based on a single-bit test. Also, each Qe value is preferably specified as a 12-bit value having the least significant bit set to 1 and having no more than four other bits set to 1. The number of Qe values in the 1/4 to 1/2 probability range is enhanced to improve coding efficiency. A decision coding parameter of preferably six bits indicates the sense of the more probable symbol (MPS) in one bit and identifies a corresponding Qe value with the remaining five bits. In addition to probability adaptation, the present invention discloses an allocation of bits in a code stream register in which preferably two spacer bits are inserted between a next byte portion (which contains a byte of data en route to a buffer) and a fractional portion which may be involved in further computation. With the two spacer bits, any code greater than or equal to Hex 'CO' which follows a Hex 'FF' byte is illegal for data and therefore provides for an escape from the code stream. The two spacer bits also reduce the number of stuff bits inserted to account for carry or borrow propagation. Encoding and decoding can be performed interchangeably by hardware or software which feature differing coding conventions.

4933883 **Probability adaptation for arithmetic coders**

William B. Pennebaker and Joan L. Mitchell

Issue Date: June 12, 1990

The present invention relates to computer apparatus and methodology for adapting the value of a probability of the occurrence of a first of two binary symbols which includes (a) maintaining a count of the number k of occurrences of the first symbol; (b) maintaining a total count of the number n of occurrences of all symbols; (c) selecting confidence limits for the probability; and (d) when the probability is outside the confidence limits, effectuating a revision in the value of the probability directed toward restoring confidence in the probability value. The number of allowed probabilities is, optionally, less than the total number of possible probabilities given the probability precision. Moreover, an approximation is employed which limits the number of probabilities to which a current probability can be changed, thereby enabling the probability adaptation to be implemented as a deterministic finite state machine.

4935882 **Probability adaptation for arithmetic coders**

William B. Pennebaker and Joan L. Mitchell

Issue Date: June 19, 1990

Apparatus and method for adapting the estimated probability of either the less likely or more likely outcome (event) of a binary decision in a sequence of binary decisions involves the up-dating of the estimated probability in response to the renormalization of an augend A. The augend A changes value with each binary decision, the size of the change depending on which of the binary events has occurred as input. Re-normalization of A occurs when the A value becomes less than a prescribed minimum value AMIN. According to the invention, there may be differing contexts in which binary decisions may occur, each context having a corresponding estimated probability value which is up-dated with binary decisions in the respective context. Also according to the invention, there may be one or more possible next values for an estimated probability in response to a given binary decision event. The selection of one of multiple possible next values in response to a given binary decision event is preferably determined based on a renormalization correlation count.

4973961 **Method and apparatus for carry-over control in arithmetic entropy coding**

Christosdoulos Chamzas and Donald L. Duttweiler
Issue Date: Nov. 27, 1990

In order to employ an output register having a finite number of stages in an arithmetic encoder, it is necessary to provide carryover control, otherwise a register having an impractically large number of stages would be required, i.e., a so-called "infinite" register. The so-called "infinite" output register is emulated by employing a counter and a finite register. To this end, a count is accumulated of sets, i.e., bytes, of consecutive prescribed logical signals of a first kind, i.e., logical 1's, being generated by an arithmetic coding register and possibly modified by a carry indication. The accumulated count is then employed to supply as an output a like number of sets including logical signals of a second kind, i.e., logical 0's, or logical signals of the first kind, i.e., logical 1's, depending on whether or not a carry would propagate through the stages of the so-called "infinite" register being emulated.

4989000 **Data string compression using arithmetic encoding with simplified probability subinterval estimation**

Dan S. Chevion, Ehud D. Karnin, and Eugeniusz Walach
Issue Date: Jan. 29, 1991

An improved method of generating a compressed representation of a source data string, each symbol of which is taken from a finite set of m+1 symbols, a_o to a_m. The method is based on an arithmetic coding procedure wherein the source data string is recursively generated as successive subintervals within a predetermined interval. The width of each subinterval is theoretically equal to the width of the previous subinterval multiplied by the probability of the current symbol. The improvement derives from approximating the width of the previous subinterval so that the approximation can be achieved by a single SHIFT and ADD operation using a suitable shift register.

5023611 **Entropy encoder/decoder including a context extractor**

Christosdoulos Chamzas and Donald L. Duttweiler

Issue Date: June 11, 1991

In entropy, e.g., arithmetic or adaptive Huffman, encoding/decoding a context based on prior symbols is needed to provide accurate predictions of symbols to be encoded/decoded. Encoding/decoding efficiency is improved by employing an adaptive context extractor. The adaptive context extractor automatically adjusts the configuration of the lag intervals used to define the context. This is realized by adaptively incorporating into the context configuration at least one lag interval found to have a "good" predictive value relative to the particular symbol stream being encoded/decoded. The context configuration adjustment is such that the at least one found lag interval is exchanged with the lag interval currently in an at least one so-called floating predictor position.

5045852 **Dynamic model selection during data compression**

Joan L. Mitchell, William B. Pennebaker, and Jorma J. Rissanen

Issue Date: Sept. 3, 1991

A system and method for maximizing data compression by optimizing model selection during coding of an input stream of data symbols, wherein at least two models are run and compared, and the model with the best coding performance for a given-size segment or block of compressed data is selected such that only its block is used in an output data stream. The best performance is determined by 1) respectively producing comparable-size blocks of compressed data from the input stream with the use of the two, or more, models and 2) selecting the model which compresses the most input data. In the preferred embodiment, respective strings of data are produced with each model from the symbol data and are coded with an adaptive arithmetic coder into the compressed data. Each block of compressed data is started by coding the decision to use the model currently being run and all models start with the arithmetic coder parameters established at the end of the preceding block. Only the compressed code stream of the best model is used in the output and that code stream has in it the overhead for selection of that model. Since the decision as to which model to run is made in the compressed data domain, i.e., the best model is chosen on the basis of which model coded the most input symbols for a given-size compressed block, rather than after coding a given number of input symbols, the model selection decision overhead scales with the compressed data. Successively selected compressed blocks are combined as an output code stream to produce an optimum output of compressed data, from input symbols, for storage or transmission.

5099440 **Probability adaptation for arithmetic coders**

William B. Pennebaker and Joan L. Mitchell

Issue Date: March 24, 1992

The present invention relates to computer apparatus and methodology for adapting the value of a probability of the occurrence of a first of two binary symbols which includes (a) maintaining a count of the number k of occurrences of the first symbol; (b) maintaining a total count of the number n of occurrences of all symbols; (c) selecting confidence limits for the probability; and (d) when the probability is outside the confidence limits, effectuating a revision in the value of the probability directed toward restoring confidence in the probability value. The number of allowed probabilities is, optionally, less than the total number of possible probabilities given the probability precision. Moreover, an approximation is employed which limits the number of probabilities to which a current probability can be changed, thereby enabling the probability adaptation to be implemented as a deterministic finite state machine.

5142283 **Arithmetic compression coding using interpolation for ambiguous symbols**

Dan S. Chevion, Ehud D. Karnin, and Eugeniusz Walach

Issue Date: Aug. 25, 1992

A method for substituting interpolated values for ambiguous symbols in an arithmetically encoded symbol string, symbols in the unencoded original string being drawn from a Markov source and a finite alphabet. Ambiguity of a symbol is defined as where its symbol occurrence statistic lies outside of a predetermined range. Interpolation involves selectively combining the occurrence statistics of symbols adjacent the ambiguous symbol and recursively and arithmetically encoding the combined statistics. The method otherwise encodes unambiguous symbols in a conventional arithmetic manner. The decoding duals are also described.

5210536 **Data compression/coding method and device for implementing said method**

Gilbert Furlan

Issue Date: May 11, 1993

A device for adaptively encoding a flow of data symbols $x_1, x_2, \ldots, x_t$ taken within a predefined symbol family, into a compressed bit string c(s), said method including so-called modelling operations (see Modelling Unit 60) whereby the numbers of times each symbol occurs in the flow are collected in a tree architecture denoting past flow symbols considered in a predefined order according to a reordering function f(t), and Relative Efficiency Counter (REC) facilities embedded in the growing tree nodes, and tracking REC variations for both optimal coding node determination and tree updating operations. The compressed bit string is then generated by an Arithmetic Coder (64) connected to said Modelling Unit (60).

5272478 **Method and apparatus for entropy coding**

James D. Allen

Issue Date: Dec. 21, 1993

The present invention provides an encoding and decoding apparatus used for the compression and expansion of data. A state machine is provided having a plurality of states. Each state has at least one transition pair. Each element of the transition pair comprises zero or more bits representative of the compact code to be output and the identification of the next state to proceed to. The transition pair reflects an output for a yes and no response associated with the probability of the data to be compacted and whether the data falls within that probability.

5307062 **Coding system**

Fumitaka Ono, Tomohiro Kimura, Masayuki Yoshida, and Shigenori Kino

Issue Date: April 26, 1994

A coding system comprises the comparing circuit which compares a magnitude of the range on the number line which is allocated to the most probability symbol with a magnitude of the fixed range on the number line which is allocated to the Less Probability Symbol. If the range allocated to the MPS is smaller than that to the LPS, and when the symbol is the MPS, the range allocated to the LPS is generated. If the range allocated to the MPS is smaller than that to the LPS, and when the symbol is the LPS, the range allocated to the MPS is generated. By the system, a coding efficiency is improved especially when a probability of occurrence of LPS (Less Probability Symbol) is approximate to 1/2.

5309381 **Probability estimation table apparatus**

Ryo Fukui

Issue Date: May 3, 1994

A probability estimation table reading apparatus includes a memory part for storing a probability estimation table, the table defining, for each address value, a switch data value, a probability estimate data value, an LPS data value indicative of a next address when a more probable symbol occurs and an MPS data value indicative of a next address when a less probable symbol occurs, and a control part for reading out each the LPS data value and MPS data value for a current address value of the stored table, so that next address values required by an arithmetic coder to perform arithmetic coding data compression, can be generated. Each of the LPS data value in the stored table stored in the memory part has been replaced by the difference between a current address value and a next address value indicated by a standard LPS data value for the current address value, and each of the MPS data value has been replaced by the difference between the current address value and a next address value indicated by a standard MPS data value for the current address value.

5311177 **Code transmitting apparatus with limited carry propagation**
Tomohiro Kimura, Fumitaka Ono, Masayuki Yoshida, and Shigenori Kino
Issue Date: May 10, 1994

Although data is transmitted with efficiency by an arithmetic encoding system, the number of carry control signals increases in proportion to the number of consecutive bits "1" s or bytes 'FF's in a conventional system. In the present invention, an arithmetic encoder 302 'detects the possibility of a carry generated during arithmetic coding operation being propagated beyond at least a predetermined number of consecutive bytes 'xFF's in a supplied arithmetic code 315. When the propagation of the carry is impossible, a carry control signal is inserted into the first 2 bits of the byte other than xFF which occurs immediately after the consecutive bytes 'xFF's so as to transmit the presence or absence of a carry. An arithmetic decoder 303 detects the continuation of at least a predetermined number of bytes 'xFF's in the arithmetic code 315, and arithmetically decodes an output value YN316 on the basis of the predicted value MPS317 of the occurrence probability of the output value YN316 to be encoded and the region width Qe of the complementary predicted value LPS. Since the number of total bits of the inserted carry control signals is reduced by this "one-time 2-bits insertion system", the total number of transmitted code bits is also reduced.

5363099 **Method and apparatus for entropy coding**
James D. Allen
Issue Date: Nov. 8, 1994

The present invention provides an encoding and decoding apparatus used for the compression and expansion of data. A state machine is provided having a plurality of states. Each state has at least one transition pair. Each element of the transition pair comprises zero or more bits representative of the compact code to be output and the identification of the next state to proceed to. The transition pair reflects an output for a yes and no response associated with the probability of the data to be compacted and whether the data falls within that probability.

5405150 **Question and answer board game with defining, spelling and synonyms**
Maria A. Loder
Issue Date: April 11, 1995

An educational game comprising a game board having a closed loop of four legs, each containing eight spaces numbered from 1-4, tokens for each player, a pair of dice, and a plurality of cards, each of which have a word and four numbered statements instructing the player what to do with that word. A player rolls the dice, moves his token into a numbered space and the player to his left reads the card and the instruction having the number corresponding to the number of his space. If the player answers correctly, he takes another turn. The winner is the first player who gets his tokens completely around the loop.

5406282 **Data coding and decoding with improved efficiency**
Yasuyuki Nomizu
Issue Date: April 11, 1995

An appearance probability forecasting device forecasts an appearance probability that a symbol described by data will be a predetermined symbol. An initial value varying device varies an initial value of the appearance probability in accordance with the type of the data. A coding device codes the data, based on the initial value of the appearance probability, the symbols actually described by the data, and the appearance probability forecast by the appearance probability forecasting device.

5414423 **Stabilization of probability estimates by conditioning on prior decisions of a given context**
William B. Pennebaker
Issue Date: May 9, 1995

A system and method involving a statistical conditioning technique that improves the coding efficiency in compression systems which have unstable statistical properties by conditioning the probability estimate for a given model context on prior decisions for that context thus enlarging the conditioning decision set. Instead of extending the model context by increasing the range of prior decisions in the pixel neighborhood, each model context is expanded into a set of two contexts which are the actual coding contexts. For a given probability estimation model context, the selection of a coding context index is done on the basis of the previous coding decision (1 or 0) for that model context. Thus, if a model context is assigned an index A, the coding context would be assigned an index 2*A +D(A)', where D(A)' is the previous (immediately preceding) binary decision for model context A; the decision of which coding context to use turns on whether the model context was used most recently for coding a 1 or a 0. More generally, each model context index A is expanded into a set of 2**N coding context indices, where N is the number of previous decisions for model context index A used in the conditioning of, for example, an arithmetic coding decision. The addressing of the context expansion can be done in any way which expands the total number of contexts by a factor of 2**N. The index for a case where the expansion is by one decision bit may be 2*A+D(A)', whereas in the case where the expansion is by two decision bits the index may be 4*A+2*D(A)'+D(A).

5546080 **Order-preserving, fast-decoding arithmetic coding arithmetic coding and compression method and apparatus**

Glen G. Langdon, Jr. and Ahmad Zandi

Issue Date: Aug. 13, 1996

An efficient, fast-decoding, order-preserving, easily implementable, length-based (L-based) arithmetic coding method, apparatus, and manufacture for an m-ary alphabet 1, . . . , i, . . . , m is provided. A coding method in accordance with the invention combines recursive division of intervals on a number line into sub-intervals whose lengths are proportional to symbol probability and which are ordered in lexical order with the constraint that probabilities be estimated as negative powers of two (1/2, 1/4, 1/8, etc.). As a consequence, the advantageous preservation of lexical order and computational efficiency are both realized. Also, a coding system in accordance with the invention is simple to implement, and high speed operation is achieved, because shifts take the place of multiplications. A coding apparatus in accordance with the invention preferably includes either a single decoding table to achieve fast decoding, or two decoding tables to achieve fast decoding as well as order preservation. The decoding process can conveniently be performed by constructing a decoding table for the C register. The C register is initialized with the leading bits of the codestring. The decoded symbol is the symbol i, i being the greatest integer that makes the C-register value greater than or equal to P(i).

A.4 Adaptive Algorithm Patents

4087788 **Data compression system**
Brian J. Johannesson
Issue Date: May 2, 1978

A data compression system is disclosed in which the left-hand boundary of a character is developed in the form of a sequence of Freeman direction codes, the codes being stored in digital form within a processor. The number of binary data bits required to define the character using different criteria is then generated and compared to determine which criteria defines the character in the minimum amount of binary data bits.

4586027 **Method and system for data compression and restoration**
Tokuhiro Tsukiyama, Yoshie Kondo, Katsuharu Kakuse, Shinpei Saba, Syoji Ozaki, and Kunihiro Itoh
Issue Date: April 29, 1986

Method of data compression and restoration wherein an input data string including repetitive data more in number than the specified value is transformed into a data string having a format including the first region where non-compressed data are placed, the second region including a datum representative of a data string section which has undergone the compression process and information indicative of the number of repetitive data, i.e., the length of the data string section, and control information inserted at the front and back of the first region indicative of the number of data included in the first region, said transformed data string being recorded on the recording medium, and, for data reproduction, the first and second regions are identified on the basis of the control information read out on the recording medium so that the compressed data string section is transformed back to the original data string in the form of repetitive data.

4633490 **Symmetrical optimized adaptive data compression/ transfer/ decompression system**

Gerald Goertzel and Joan L. Mitchell
Issue Date: Dec. 30, 1986

Data compression for transfer (storage or communication) by a continuously adaptive probability decision model, closely approaches the compression entropy limit. Sender and receiver perform symmetrical compression/decompression of binary decision n according to probabilities calculated independently from the transfer sequence of $1 \ldots n-1$ binary decisions. Sender and receiver dynamically adapt the model probabilities, as a cumulative function of previously presented decisions, for optimal compression/decompression. Adaptive models for sender and receiver are symmetrical, to preserve data identity; transfer optimization is the intent. The source model includes a state generator and an adaptive probability generator, which dynamically modify the coding of decisions according to state, probability and bit signals, and adapt for the next decision. The system calculates probability history for all decisions, including the current decision, but uses probability history for decision $n-1$ (the penultimately current decision) for encoding decision n (the dynamically current decision). The system, separately at source and destination, reconfigures the compression/expansion algorithm, up to decision $n-1$, codes each decision in the data stream optimally, according to its own character in relation to the calculated probability history, and dynamically optimizes the current decision according to the transfer optimum of the data stream previously transferred. The receiver operates symmetrically to the sender. Sender and receiver adapt identically, and adapt to the same decision sequence, so that their dynamically reconfigured compression-expansion algorithms remain symmetrical–even though the actual algorithms may change with each decision as a function of dynamic changes in probability history.

4941092 **Signal processing method for determining base sequence of nucleic acid**

Makoto Hara, Masakazu Hashiue, and Kazuyoshi Tanaka
Issue Date: July 10, 1990

A signal processing method for determining base sequence of nucleic acids by subjecting digital signals to signal processing, said digital signals corresponding to an autoradiograph of plural resolved rows which are formed by resolving a mixture of base-specific DNA fragments or base-specific RNA fragments labeled with a radioactive element in one-dimensional direction on a support medium, which comprises steps of: (1) detecting at least two bands continuously in the lower part of each resolved row and numbering the bands consecutively in order from the lower end; (2) obtaining correlation of a distance between the detected bands in the resolving direction with the band's number and predicting positions of undetected bands in the resolving direction from the correlation; (3) detecting at least one band on the resolved rows on the basis of the predicted positions and numbering the bands consecutively; (4) obtaining the correlation of the distance between the bands with the band's number for the already detected bands including the band newly detected in the step (3), and predicting positions of undetected bands from the correlation; and (5) repeating in order the steps (3) and (4) to thereby detect all bands on the resolved rows.

4872009 **Method and apparatus for data compression and restoration**

Tokuhiro Tsukiyama, Hiroshi Yashiki, and Osamu Hirose
Issue Date: Oct. 3, 1989

A method of data compression for recording data on a recording medium such as a magentic tape, a method of data restoration for data which has been compressed for recording, and an apparatus of data compression and restoration prescribe the data to be compressed based on type or value and encode the compression object data, thereby reducing the number of bits necessary to indicate the compression object data. Compression is implemented only for consecutive data fewer in number of repeating consecutive bytes than a certain number, thereby reducing the number of bits necessary to indicate the number of bytes of the consecutive data. A compression mark indicative of compression is appended to the compressed data, consisting of data made by encoding the compression object consecutive data and data indicating the number of bytes of the data, either at the front or rear of the compressed data, whereby an input data string can be compressed drastically and compressed data, even including errors, can be restored.

4876541 **Stem for dynamically compressing and decompressing electronic data**

James A. Storer
Issue Date: Oct. 24, 1989

A data compression system for encoding and decoding textual data, including an encoder for encoding the data and for a decoder for decoding the encoded data. Both encoder and decoder have dictionaries for storing frequently-appearing strings of characters. Each string is identified by a unique pointer. The input data stream is parsed and matched with strings in the encoder dictionary using a novel matching algorithm. The pointer associated with the matched string is then transmitted to a remote location for storage or decoding. Thereafter, using a novel update algorithm the encoder dictionary is updated to include new strings of data based on the matched string of data. If required, a novel deletion algorithm is employed to remove infrequently used strings of data to provide room for the newly generated strings of data. The strings of data may be arranged using a modified least recently used queue. The decoder matches each unique pointer in the stream of compressed input data with a corresponding pointer in the decoder dictionary. The decoder then transmits the string of character data associated with the matched pointer, thereby providing textual data in original, uncompressed form. Thereafter, using the novel update and deletion algorithms, new strings of data are added to, and old strings of data are deleted from, the decoder dictionary, so as to ensure both encoder and decoder dictionaries contain identical strings of data.

5003307 **Data compression apparatus with shift register search means**

Douglas L. Whiting and Glen A. George
Issue Date: March 26, 1991

An apparatus and method are disclosed for converting an input data character stream into a variable length encoded data stream in a data compression system. The data compression system includes a shift register means. The shift register means has a plurality of entries and each entry of the shift register means is for storing a data character of the input data stream. The method for converting the input data character stream includes the following steps. Performing a search in the shift register means for a data string which matches the input data string. The step for performing the search includes the steps of broadcasting each input data character of the input data stream to each entry of the shift register means and comparing each input data character simultaneously with the previously stored contents of each entry of said shift register means. If the matching data string is found within the shift register means, the next step includes encoding the longest matching data string by appending to the encoded data stream a tag indicating the matching data string and a string substitution code. If the matching data string is not found within the shift register means, the next step includes encoding the first character of the input data string by appending to the encoded data stream a raw data tag and the first character of the input data string.

A.5 Grammar Algorithm Patents

3976844 **Data communication system for transmitting data in compressed form**

Bernard K. Betz

Issue Date: Aug. 24, 1976

Characters which repeat between successive groups or lines on a page, are, except for an initial transmission, not transmitted, thereby decreasing the time for data transmission. Only the non-repeat characters between the successive groups or lines are transmitted along with a "map" (i.e. a sequence of numbers) indicative of the format of the repeat and non-repeat characters of the present group, or line, so as to enable a receiving station to reassemble the compressed data. A new map need not be sent, thereby further decreasing transmission time, if the transmission using an old map which has been previously transmitted compares favorably with the transmission using the new one.

4295125 **Method and means for pipeline decoding of the high to low order pairwise combined digits of a decodable set of relatively shifted finite number of strings**

Glen G. Langdon, Jr.

Issue Date: Oct. 13, 1981

An apparatus for ensuring continuous flow through a pipeline processor as it relates to the serial decoding of FIFO Rissanen/Langdon arithmetic string code of binary sources. The pipeline decoder includes a processor (11, 23) and a finite state machine (21, FSM) in interactive signal relation. The processor generates output binary source signals (18), status signals (WASMPS, 31) and K component/K candidate next integer-valued control parameters (L0, k0; L1, k1; 25). These signals and parameters are generated in response to the concurrent application of one bit from successive arithmetic code bits, a K component present integer-value control parameter (52) and K component vector representation (T, TA) of the present internal state (51) of the associated finite state machine (FSM). The FSM makes a K-way selection from K candidate next internal states and K candidate next control parameters. This selection uses no more than $K^2 + K$ computations. The selected signals are then applied to the processor in a predetermined displaced time relation to the present signals in the processor. As a consequence, this system takes advantage of the multi-state or "memory" capability of an FSM in order to control the inter-symbol influence and facilitate synchronous multi-stage pipeline decoding.

4463342 **Method and means for carry-over control in the high order to low order pairwise combining of digits of a decodable set of relatively shifted finite number strings**
Glen G. Langdon, Jr. and Jorma J. Rissanen
Issue Date: July 31, 1984

Carry-over control in strings resulting from the high to low order combining of two binary number strings is obtained through the insertion of a control character within the resultant string after detecting a run of consecutive 1's. Upon subsequent accessing and decomposition of the resultant string, the control character causes string decomposition to operate for a number of cycles in a carry correction mode. If the control character indicates that a carry has rippled through the n lesser significant positions of the resultant string, then upon decomposition, those "n" consecutive 1's are changed to 0's, and a 1 is added to the least significant position in the string preceding the control character. If the control character indicates no carry occurrence, then it is merely deleted from the string. The control of carries in this manner permits the generation of arithmetic string compression code sequences in an instantaneous FIFO pattern with only a modest reduction of compression efficiency. Relatedly, the encoder and carry suppressor is shown in FIG. 2[2], while the carry corrector and decoder is shown in FIG. 4.

A.6 Quantization Algorithm Patents

4366551 **Associative memory search system**
Klaus E. Holtz
Issue Date: Dec. 28, 1982

A knowledge storage and retrieval method is performed to result in a single address number of a storage region which represents a large quantity of information. A first character of a data sequence and a starting number are directed into a two-position buffer to form an input matrix. The storage locations of the storage region are addressed to see if the matrix is found in the storage region. If the matrix is found, the address number of the storage location of the storage region at which it is found is then stored in the second position of the buffer region. Then the

[2] See original patent for figures.

next data character is directed into the first position of the buffer region and the searching step is repeated. If the matrix in the buffer region is not found in the storage region, the matrix is stored in a free storage location of the storage region and the address number of the free storage location is directed into the second position of the buffer. The next data character is moved into the first position and the searching step is repeated. Eventually, all of the data characters of a data sequence will be stored and the address number representing a storage location of the last-to-be stored data character will represent all of the data characters of the data sequence that were stored.

A.7 Image Algorithm Patents

4622545 **Method and apparatus for image compression and manipulation**
William D. Atkinson
Issue Date: Nov. 11, 1986

Apparatus and methods are disclosed which are most advantageously used in conjunction with a digital computer to provide improved graphics capability. These techniques permit the representation and manipulation of any arbitrarily shaped image in terms of "inversion points". Inversion points defining a region are sorted and stored such that the region shape may be regenerated at a later time from the inversion points. Means are provided to compare existing regions and new regions to be displayed, and region operators are provided to specify a precedence between the existing and new regions. Thus, new regions are appropriately "clipped" such that only portions of a new region may actually be displayed to achieve the desired graphic representation.

4667649 **Archery bow**
Stanley A. Humphrey
Issue Date: May 26, 1987

An archery bow having a vertically elongated handle, upper and lower bow limbs that are pivotally connected to the handle intermediate opposite ends of the limbs, a pair of levers pivotally connected to the opposite ends of the handle, swivel arms pivotally mounted to the handle, first and second idler pulleys mounted on the one ends of the swivel arms. A pulley wheel is mounted to each of the remote ends of the limbs. The bowstring connects to and extends from each lever to the respective adjacent idler wheels. The bowstring is reversely bent over each idler wheel and extends therefrom to the adjacent pulley wheels on the limbs remote ends whereby the bowstring is reversely bent thereof. A cable joins and extends between each adjacent lever and limb end portion whereby as the bowstring is drawn the idler wheels are moved toward one another to pivot the swivel arms such that the cables draw the adjacent ends of the limbs to apply bowing forces thereto while bowing forces are also applied to the remote ends of the limbs through the wheels mounted thereon.

4809350 **Data compression system**

Yair Shimoni and Ron Niv
Issue Date: Feb. 28, 1989

A data compression system for use in processing diagnostic image data which uses a predictor to predict future data. The actual data is subtracted from the predicted value to obtain data related difference values. The difference values are coded by coding the most prevalent difference by a code other than the least bit code but further coding the most prevalent number by string length coding, with the most common being the coded with a least bit code, and also coding the next most common difference with a least bit code and the subsequent most prevalent differences with sequential least bit codes to provide compression ratios of over 3:1. A procedure for finding the best predictor in noisy data is provided.

4941193 **Methods and apparatus for image compression by iterated function system**

Michael F. Barnsley and Alan D. Sloan
Issue Date: July 10, 1990

A method and apparatus for obtaining highly compressed images employing an iterated function system (IFS). An original input or target image is subdivided into regions having similar characteristics. Contractive copies or maps of a particular region, which are the results of affine transformations to the region, are generated and tiled with respect to the input image until the entire region is covered and a collage is formed. Each region is processed in like manner. The affine transformation coefficients or IFS codes completely represent the input image, and are stored or transmitted. To generate an image from the IFS codes, a decoding system is disclosed. One disclosed method involves a chaotic dynamical system. A random iteration of the IFS codes is performed until an attractor, which is the target image, emerges and stabilizes. Another disclosed deterministic method repeatedly and successively applies the IFS codes to an arbitrary starting image until the attractor emerges. Also disclosed are various methods for representing and compressing the color information of an image, including a method for employing an additional spatial dimension in the mappings and a method for employing an arbitrary probabilistic measure for the color rendering.

4943869 **Compression method for dot image data**
Hiroshi Horikawa, Hitoshi Urabe, and Katsutoshi Yako
Issue Date: July 24, 1990

This invention method performs data compression for dot image data by utilizing the redundancy based on spacial correlation of images and correlation of dot patterns. More particularly, a threshold matrix is rearranged in array according to a predetermined reference, the threshold matrix being used to convert input image data into dot data, so as to output the same logical values continuously from the top and/or from the end of dot data array, and simultaneously dot data are grouped in the unit of a predetermined bits so as to count the number of units having the same and continuous logical values from the top and/or from the end to thereby effectively compress data of dot images by a perfect reproduction system or a non-perfect reproduction system.

A.8 Fractal Algorithm Patents

5065447 **Method and apparatus for processing digital data**
Michael F. Barnsley and Alan D. Sloan
Issue Date: Nov. 12, 1991

Digital image data is automatically processed by dividing stored image data into domain blocks and range blocks. The range blocks are subjected to processes such as a shrinking process to obtain a mapped range blocks. Then, for each domain block, the mapped range block which is most similar to the domain block is determined, and the address of that range block and the processes the block was subjected to are combined as an identifier which is appended to a list of identifiers for other domain blocks. The list of identifiers for all domain blocks is called a fractal transform and constitutes a compressed representation of the input image. To decompress the fractal transform and recover the input image, an arbitrary input image is formed into range blocks and the range blocks processed in a manner specified by the identifiers to form a representation of the original input image.

5347600 **Method and apparatus for compression and decompression of digital image data**

Michael F. Barnsley and Alan D. Sloan

Issue Date: Sept. 13, 1994

Digital image data is automatically processed by dividing stored image data into domain blocks and range blocks. The range blocks are subjected to processes such as a shrinking process to obtain mapped range blocks. The range blocks or domain blocks may also be processed by processes such as affine transforms. Then, for each domain block, the mapped range block which is most similar to the domain block is determined, and the address of that range block and the processes the blocks were subjected to are combined as an identifier which is appended to a list of identifiers for other domain blocks. The list of identifiers for all domain blocks is called a fractal transform and constitutes a compressed representation of the input image. To decompress the fractal transform and recover the input image, an arbitrary input image is formed into range blocks and the range blocks processed in a manner specified by the identifiers to form a representation of the original input image.

5384867 **Fractal transform compression board**

Michael F. Barnsley, Alan D. Sloan, John H. Elton, Charles S. Moreman, and Guy A. Primiano

Issue Date: Jan. 24, 1995

Digital image data compression apparatus includes a controller circuit for receiving digital image data and for processing the image data into blocks. The controller circuit supplies processed image data to a plurality of transform circuits and to a feeder circuit. The transform circuits receive data from the controller circuit and the feeder circuit, and provide parallel processing to compare blocks of image data and generate fractal transform values representing the image data in a compressed form.

5416856 **Method of encoding a digital image using iterated image transformations to form an eventually contractive map**

Everett W. Jacobs, Roger D. Boss, and Yuval Fisher

Issue Date: May 16, 1995

A method is provided for encoding an image in an iterated transformation image compression system. The image, represented by an array of pixels, is partitioned into ranges and domains. A transformation is generated for each domain such that no transformation is constrained to be contractive. Each domain's transformation is optimized in terms of the intensity scaling and offset coefficients. A domain and corresponding optimized transformation is selected that minimizes transformation error data. These steps are repeated for a chosen plurality of the ranges that form a non-overlapping tiling of the image and such that the corresponding optimized transformations associated with the chosen plurality form an eventually contractive map. Information that identifies each of the chosen plurality of the ranges and the selected domains and corresponding optimized transformations is then stored in an addressable memory as an encoding of the image.

5430812 **Fractal transform compression board**

Michael F. Barnsley, Alan D. Sloan, John H. Elton, Charles S. Moreman, and Guy A. Primiano

Issue Date: July 4, 1995

Digital image data compression apparatus includes a controller circuit for receiving digital image data and for processing the image data into blocks. The controller circuit supplies processed image data to a plurality of transform circuits and to a feeder circuit. The transform circuits receive data from the controller circuit and the feeder circuit, and provide parallel processing to compare blocks of image data and generate fractal transform values representing the image data in a compressed form.

A.9 Other Patents

4758899 **Data compression control device**

Tokuhiro Tsukiyama

Issue Date: July 19, 1988

A control device has a data compression control circuit for compressing data and a data buffer for temporarily storing the data to be written on a recording medium. The data transferred from a CPU is compressed and then recorded on the recording medium. Further provided is a buffer discriminating circuit for discriminating the quantity of data stored in the data buffer to control the data compression control circuit so that the data compressing operation and the operation of recording the data on the recording medium may be conducted in parallel while the amount of transferred data in the data buffer is prevented from becoming less than a predetermined quantity.

5533051 **Method for data compression**

David C. James

Issue Date: July 2, 1996

Methods for compressing data including methods for compressing highly randomized data are disclosed. Nibble encode, distribution encode, and direct bit encode methods are disclosed for compressing data which is not highly randomized. A randomized data compression routine is also disclosed and is very effective for compressing data which is highly randomized. All of the compression methods disclosed operate on a bit level and accordingly are insensitive to the nature or origination of the data sought to be compressed. Accordingly, the methods of the present invention are universally applicable to any form of data regardless of its source of origination.

Appendix B

Bibliography

Compression is an active and important field of study, and it has been that way for some time. The earliest papers date to the vacuum tube era when scientists at Bell Labs were first trying to understand the concept of "information." Today, there are many conferences, including one major one held each winter. The worlds of video and audio encoding are now separate industries with their own cliques, conferences, and agendas.

This book could not be written without relying on the work of many of the researchers in the field. Unfortunately, I have the feeling that many of the details in the book were once discovered by someone and published with great excitement only to be absorbed into the general body of knowledge. If you know of any text that went uncited, please bring it to my attention. I will include it in an errata sheet on my Website.

The rest of this appendix contains a list of the important books and papers cited. This list can be a bit barren and imposing at times. For this reason, I offer these suggestions to researchers:

- The Data Compression Conference held each year is a good source of the latest research. The procedings are published by the I.E.E.E. Computer Society, (for instance, [var98]).
- Several of the general introductions are quite good. James Storer's book is a great introduction to the statistical and dictionary-based approaches [Sto88].
- Mark Nelson's book is a good introduction to the topic, and it comes with

some good examples of source code [Nel92].

- The book by K. R. Rao and J. J. Hwang is a good introduction to image coding [RH96].

[Abr63] Norman Abramson. *Information Theory and Coding*. McGraw-Hill, New York, NY, 1963.

[Aik88] A. Aiken. Compaction-based parallelization. Report TR 88-922, Dept. of Computer Science, Cornell Univerisity, Ithaca, NY, June 1988.

[AN91] A. Aiken and A. Nicolau. A realistic resource-constrained software pipelining algorithm. In A. Nicolau *et al.*, editors, *Advances in languages and compilers for parallel processing*, pages 274–290. Pitman/MIT Press, London, 1991.

[ano86] anon. *Postscript Language Reference Manual*. Addison-Wesley, Reading, MA, 1986.

[ANR74] N. Ahmed, T. Natarajan, and K. R. Rao. Discrete cosine transform. *IEEE Transactions on Computers*, C-23(1):90–93, January 1974.

[AT89] S. Abenda and G. Turchetti. Inverse problems of fractal sets on the real line via the moment method. *Nuovo Cimento*, 104B:213–227, 1989.

[ATD92] S. Abenda, G. Turchetti, and S. Demko. Local moments and inverse problem for fractal measures. *Inverse Problems*, 8:739–750, 1992.

[Bar88] Michael F. Barnsley. Fractal modelling of real world images. In Heinz-Otto Peitgen and Dietmar Saupe, editors, *The Science of Fractal Images*, chapter 5, pages 219–239. Springer-Verlag, 1988.

[Bar93] Michael F. Barnsley. *Fractals Everywhere*. Academic Press, Cambridge, MA, 2nd edition, 1993.

[BD86] Michael F. Barnsley and Stephen G. Demko, editors. *Chaotic Dynamics and Fractals*, volume 2 of *Notes and reports in mathematics in science and engineering*. Academic Press, London, 1 edition, 1986.

[BG90] N. Bellare and S. Goldwasser. New paradigms for digital signatures and message authentication based on non interactive zero knowledge proofs. In *Advances in Cryptology- CRYPTO '89 Proceedings*. Springer-Verlag, 1990.

[BM96] P. Bird and T. Mudge. An instruction stream compression technique. Technical Report CSE-TR-319-96, EECS Department, University of Michigan, Ann Arbor, MI, November 1996.

[BMK93] Zachi Barahav, David Malah, and Ehud Karnin. Hierarchical interpretation of fractal image coding and its application to fast decoding. In *Proceedings of the IEEE International Conference on Digital Signal Processing*, pages 190–195, Nicosia, Cyprus, July 1993.

[BMK95] Zachi Barahav, David Malah, and Ehud Karnin. Hierarchical interpretation of fractal image coding and its applications. In Yuval Fisher, editor, *Fractal Image Compression: Theory and Application*, Chapter 5, pages 91–117. Springer-Verlag, New York, NY, 1995.

[Bog94a] Alexandru Bogdan. Multiscale fractal image coding and the two-scale difference equation. Technical Report CU/CTR/TR 358-94-05, Columbia University, New York, March 1994.

[Bog94b] Alexandru Bogdan. Multiscale (inter/intra frame) fractal video coding. In *Proc. IEEE International Conference on Image Processing. ICIP '94*, Austin, TX, November 1994.

[Bon96] Robert Bonneau. Multiresolution transform and its application to image coding. In Rashid Ansari and Mark J. Smith, editors, *Visual Communications and Image Processing '96*, volume 2727 of *SPIE Proceedings*, pages 560–567, 1996.

[BZM98a] Ali Bilgin, George Zweig, and Michael W. Marcellin. Efficient lossless coding of medical image volumes using reversible integer wavelet transforms. In *Proceedings DCC'98 (IEEE Data Compression Conference)*, 1998. Preprint.

[BZM98b] Ali Bilgin, George Zweig, and Michael W. Marcellin. Lossless medical image compression using three-dimensional integer wavelet transforms. *IEEE Transactions on Medical Imaging*, 1998. Preprint.

[CBM96] I. Chen, P. Bird, and T. Mudge. The impact of instruction compression on i-cache performance. Technical Report CSE-TR-330-97, EECS Department, University of Michigan, 1996.

[CBSC94] Charilaos A. Christopoulos, Jan G. Bormans, Athanasios N. Skodras, and Jan P. Cornelis. Efficient computation of the two-dimensional fast cosine transform. In David P. Casasent and Andrew G. Tescher, editors, *Hybrid Image and Signal Processing IV*, volume 2238 of *SPIE Proceedings*, pages 229–237, Orlando, FL, April 1994.

[CFMV92] Carlos A. Cabrelli, Bruno Forte, Ursula M. Molter, and Edward R. Vrscay. Iterated fuzzy set systems: A new approach to the inverse problem for fractals and other sets. *Journal of Mathematical Analysis and Applications*, 171(1):79–100, November 1992.

[CG91] Vincent Cate and Thomas Gross. Integration of compression and caching for a two-level file system. In *Proc. Symp. on Architectural Support for Programming Languages and Operating Systems*, pages 200–211, 1991.

[CGO94] Pamela C. Cosman, Robert M. Gray, and Richard A. Olshen. Evaluating quality of compressed medical images: SNR, subjective rating, and diagnostic accuracy. *Proceedings of the IEEE*, 82(6):919–932, June 1994.

[CGV96] Pamela C. Cosman, Robert M. Gray, and Martin Vetterli. Vector quantization of image subbands: A survey. *IEEE Transactions on Image Processing*, 5(2):202–225, February 1996.

[Che97] I. Chen. *Enhancing Instruction Fetching Mechanism Using Data Compression*. Ph.D. thesis, University of Michigan, 1997.

[Chu92a] Charles K. Chui. *An Introduction to Wavelets*, volume 1 of *Wavelet Analysis and its Applications*. Academic Press, San Diego, CA, 1992.

[Chu92b] Charles K. Chui, editor. *Wavelets: A Tutorial in Theory and Applications*, volume 2 of *Wavelet Analysis and its Applications*. Academic Press, San Diego, CA, 1992.

[Chu97] Charles K. Chui. *Wavelets: A Mathematical Tool for Signal Analysis*. Society for Industrial and Applied Mathematics, Philadelphia, PA, 1997.

[CL91] C. C. Chang and C. H. Lin. An id-based signature scheme based upon Rabin's public key cryptosystem. *Proceedings of the 25th Annual 1991 IEEE International Carnahan Conference on Security Technology*, Oct 1 –3 1991.

[CR95] Albert Cohen and Robert D. Ryan. *Wavelets and Multiscale Signal Processing*. Number 11 in Applied Mathematics and Mathematical Computation. Chapman & Hall, London, England, 1995.

[CSF77] Wen-Hsiung Chen, C. Harrison Smith, and S. C. Fralick. A fast computational algorithm for the discrete cosine transform. *IEEE Transaction on Communications*, COM-25(9):1004–1009, September 1977.

[CT91] Thomas M. Cover and Joy A. Thomas. *Elements of Information Theory*. Wiley Series in Telecommunications. John Wiley & Sons, New York, NY, 1991.

[CZ97] Bing Cheng and Xiaokun Zhu. Multiresolution approximation of fractal transform. Submitted to *Signal Processing*, 1997. Preprint.

[DACB96] Franck Davoine, Marc Antonini, Jean-Marc Chassery, and Michel Barlaud. Fractal image compression based on Delaunay triangulation and vector quantization. *IEEE Transactions on Image Processing*, 5(2):338–346, February 1996.

[Dau92] Ingrid Daubechies. *Ten Lectures on Wavelets*. Society for Industrial and Applied Mathematics, Philadelphia, PA, 1992.

[DBC93] F. Davoine, E. Bertin, and J-M. Chassery. From rigidity to adaptive tesselations for fractal image compression: Comparative studies. In *IEEE 8th Workshop on Image and Multidimensional Signal Processing*, September 1993.

[DC94] F. Davoine and J-M. Chassery. Adaptive delaunay triangulation for attractor image coding. In *12th International Conference on Pattern Recognition*, October 1994.

[DSC95] Franck Davoine, John Svensson, and Jean-Marc Chassery. A mixed triangular and quadrilateral partition for fractal image coding. In *Proceedings ICIP-95 (IEEE International Conference on Image Processing)*, volume III, pages 284–287, Washington, D.C., October 1995.

[EFE+97] Jens Ernst, Christopher W. Fraser, William Evans, Steven Lucco, and Todd A. Proebsting. Code compression. In *Proc. Conf. on Programming Languages Design and Implementation*, pages 358–365, June 1997.

[Fal90] Kenneth Falconer. *Fractal Geometry: Mathematical Foundations and Applications*. John Wiley & Sons, Chichester, England, 1990.

[FK96] M. Franz and T. Kistler. Slim binaries. Technical Report No. 96-24, Department of Information and Computer Science, University of California, Irvine, 1996.

[FMW84] C. Fraser, E. Myers, and A. Wendt. Analyzing and compressing assembly code. In *Proc. of the ACM SIGPLAN Symposium on Compiler Construction*, volume 19, pages 117–121, June 1984.

[FP95] Christopher W. Fraser and Todd A. Proebsting. Custom instruction sets for code compression. *Not published*, 1995.

[Fra94] M. Franz. *Code-Generation On-The-Fly: A Key to Portable Software*. Ph.D. thesis, Swiss Federal Institute of Technology, 1994.

[Fra96] M. Franz. Adaptive compression of syntax trees and iterative dynamic code optimization: Two basic technologies for mobile-object systems. Technical Report No. 96-04, Department of Information and Computer Science, University of CA, Irvine, 1996.

[Fra99] Christopher W. Fraser. Automatic inference of models for statistical code compression. In *Proc. Conf. on Programming Languages Design and Implementation*, 1999.

[FV94a] Bruno Forte and Edward Vrscay. Solving the inverse problem for function/image approximation using iterated function systems: I theoretical basis. *Fractals*, 2(3):325–334, 1994.

[FV94b] Bruno Forte and Edward Vrscay. Solving the inverse problem for function/image approximation using iterated function systems: II algorithms and computations. *Fractals*, 2(3):335–346, 1994.

[FW86] Christopher W. Fraser and Alan L. Wendt. Integrating code generation and optimization. *SIGPLAN Notices*, 21(7):242–248, June 1986.

[Gal68] Robert G. Gallager. *Information Theory and Reliable Communication*. John Wiley & Sons, New York, NY, 1968.

[Ham80] Richard W. Hamming. *Coding and Information Theory*. Prentice-Hall, London, England, 1980.

[Ham97] Raouf Hamzaoui. Codebook clustering by self-organizing maps for fractal image compression. *Fractals*, 5 (Supplementary Issue):27–38, April 1997.

[Har94] John C. Hart. Fractal image compression and the inverse problem of recurrent iterated function systems. *IEEE Computer Graphics and Applications*, July 1994.

[HCF97] John C. Hart, Wayne O. Cochran, and Patrick J. Flynn. Similarity hashing: A computer vision solution to the inverse problem of linear fractals. *Fractals*, 5 (Supplementary Issue):39–50, April 1997.

[He96] Yu He. *Mathematica: Wavelet Explorer*. Wolfram Research, Champaign, Illinois, June 1996.

[Hec82] Paul Heckbert. Color image quantization for frame buffer display. In *Proceedings of SIGGRAPH 82*, 1982.

[HM90] C. Handy and Giorgio Mantica. Inverse problems in fractal construction: Moment method solution. *Physica D*, 43:17–36, 1990.

[Hol91] Scott A. Hollatz. Digital image compression with two-dimensional affine fractal interpolation functions. Technical Report 91-2, Department of Mathematics and Statistics, University of Minnesota - Duluth, 1991.

[Hou87] Hsieh S. Hou. A fast recursive algorithm for computing the discrete cosine transform. *IEEE Transactions on Acoustics, Speech, and Signal Processing*, ASSP-35(10):1455–1461, October 1987.

[HS93] Bernd Hürtgen and Christoph Stiller. Fast hierarchical codebook search for fractal coding of still images. In *SPIE Visual Communication and PACS for Medical Applications*, pages 397–408, Berlin, Germany, 1993.

[HU79] John E. Hopcroft and Jeffrey D. Ullman. *Introduction to Automata Theory, Languages, and Computation*. Addison-Wesley, Reading, MA, 1979.

[Huf52] David A. Huffman. A method for the construction of minimum-redundancy codes. *Proceedings of the IRE*, 40(9):1098–1101, September 1952.

[JS97] Mauricio Breternitz Jr. and Roger Smith. Enhanced compression techniques to symplify program decompression and execution. In *Proc. Int'l Conf. on Computer Design*, pages 170–176, October 1997.

[Kai94] Gerald Kaiser. *A Friendly Guide to Wavelets*. Birkhäuser, Boston, MA, 1994.

[KLH98] Jason Knipe, Xiabo Li, and Bin Han. An improved lattice vector quantization scheme for wavelet compression. *IEEE Transactions on Signal Processing*, 46(1):239–243, January 1998.

[Kom95] John Kominek. Codebook reduction in fractal image compression. In Robert L. Stevenson, Alexander I. Drukarev, and Thomas R. Gardos, editors, *Still-Image Compression II*, volume 2669 of *SPIE Proceedings*, pages 33–41, San Jose, CA, January 1995.

[KW94] M. Kozuch and A. Wolfe. Compression of embedded system programs. In *Proc. Int'l Conf. on Computer Design*, 1994.

[KW95] Michael Kozuch and Andrew Wolfe. Performance analysis of the compressed code RISC processor. Technical Report CE-A95-2, Princeton University Computer Engineering, 1995.

[Lan84] Glen G. Langdon, Jr. An introduction to arithmetic coding. *IBM Journal of Research and Development*, 28(2):135–149, March 1984.

[LBCM97] Charles Lefurgy, Peter Bird, I-Cheng Chen, and Trevor Mudge. Improving code density using compression techniques. In *Proceedings of the 30th Annual International Symposium on Microarchitecture*, pages 194–203, December 1997.

[LDK95a] Stan Liao, S. Devadas, and K Keutzer. Code density optimization for embedded DSP processors using data compression techniques. In *Proc. Conf. on Advanced Research in VLSI*, 1995.

[LDK+95b] Stan Liao, Srinivas Devadas, Kurt Keutzer, Steve Tjiang, and Albert Wang. Code optimization techniques for embedded DSP microprocessors. In *Proc. ACM/IEEE Design Automation Conference*, 1995.

[LDK+95c] Stan Liao, Srinivas Devadas, Kurt Keutzer, Steve Tjiang, and Albert Wang. Storage assignment to decrease code size. *SIGPLAN Notices*, 30(6):186–195, June 1995.

[LDK+96] Stan Liao, Srinivas Devadas, Kurt Keutzer, Steven Tjiang, and Albert Wang. Storage assignment to decrease code size. *ACM Transactions on Programming Languages and Systems*, 18(3):235–253, May 1996.

[Lia96] Stan Liao. *Code Generation and Optimization for Embedded Digital Signal Processors*. Ph.D. thesis, Massachusetts Institute of Technology, June 1996.

[LM98] Charles Lefurgy and Trevor Mudge. Code compression for dsp. Technical Report CSE-TR-380-98, University of Michigan, Ann Arbor, November 1998.

[LMR97] A. K. Louis, P. Maaß, and A. Rieder. *Wavelets: Theory and Applications*. John Wiley & Sons, Chichester, England, 1997.

[LØ95] Skjalg Lepsøy and Geir E. Øien. Fast attractor image encoding by adaptive codebook clustering. In Yuval Fisher, editor, *Fractal Image Compression: Theory and Application*, chapter 9, pages 177–197. Springer-Verlag, New York, NY, 1995.

[LRO97] Scott M. Lopresto, Kannan Ramchandran, and Michael T. Orchard. Image coding based on mixture modeling of wavelet coefficients and a fast estimation-quantization framework. In *Proceedings DCC'97 (IEEE Data Compression Conference)*, 1997.

[Man77] Benoit B. Mandelbrot. *The Fractal Geometry of Nature*. W.H. Freeman & Company, New York, NY, 1977.

[Mey92] Yves Meyer. *Wavelets and Operators*. Cambridge University Press, Cambridge, 1992.

[Mey93] Yves Meyer. *Wavelets: Algorithms and Applications*. Society for Industrial and Applied Mathematics, Philadelphia, PA, 1993.

[MH91] David S. Mazel and Monson H. Hayes. Hidden-variable fractal interpolation of discrete sequences. In *IEEE Proceedings of ICASSP*, pages 3393–3396, May 1991.

[MS93] Donald M. Monro and Barry G. Sherlock. Optimum DCT quantization. In James A. Storer and Martin Cohn, editors, *Proceedings DCC'93 (IEEE Data Compression Conference)*, pages 188–194, Snowbird, UT, April 1993.

[Nel92] Mark Nelson. *The Data Compression Book*. M&T Books, New York, NY, 1992.

[NMW97] Craig G. Nevill-Manning and Ian H. Witten. Phrase hierarchy inference and compression in bounded space. In *Proceedings of Data Compression Conference 1997*, pages 3–11, Los Alamitos, CA; Piscataway, NJ; Brussels; Tokyo, March 1997. IEEE Computer Society.

[NMW98] Craig G. Nevill-Manning and Ian H. Witten. Phrase hierarchy inference and compression in bounded space. In *Proceedings of Data Compression Conference 1998*, pages 179–188, Los Alamitos, CA; Piscataway, NJ; Brussels; Tokyo, April 1998. IEEE Computer Society.

[NP78] Madihally J. Narasimha and Allen M. Peterson. On the computation of the discrete cosine transform. *IEEE Transaction on Communications*, COM-26(6):934–936, June 1978.

[PM93] William B. Pennebaker and Joan L. Mitchell. *JPEG Still Image Data Compression Standard*. Van Nostrand Reinhold, New York, NY, 1993.

[PMN96] Shang H. Park, Hyeon J. Moon, and Nasser M. Nasrabadi. Subband image coding using block-zero tree coding and vector quantization. In *Proceedings ICASSP-96 (IEEE International Conference on Acoustics, Speech and Signal Processing)*, volume 4, pages 2054–2057, 1996.

[RC94] R. Rinaldo and G. Calvagno. An image coding scheme using block prediction of the pyramid subband decomposition. In *IEEE International Conference on Image Processing (ICIP '94)*, Austin, Texas, November 1994.

[RH96] K.R. Rao and J.J. Hwang. *Techniques & Standards for Image, Video & Audio Coding*. Prentice Hall, Upper Saddle River, NJ, 1996.

[RL79] J. Rissanen and Glen G. Langdon, Jr. Arithmetic coding. *IBM Journal of Research and Development*, 23(2):149–162, March 1979.

[RL81] Jorma Rissanen and Glen G. Langdon, Jr. Universal modeling and coding. *IEEE Transactions on Information Theory*, IT-27(1):12–23, January 1981.

[RL95] Viresh Ratnakar and Miron Livny. RD-OPT: An efficient algorithm for optimizing DCT quantization tables. In *Proceedings DCC'95 (IEEE Data Compression Conference)*, pages 332–341, 1995.

[Saf94] Robert J. Safranek. A JPEG compliant encoder utilizing perceptually based quantization. In Bernice E. Rogowitz and Jan P. Allebach, editors, *Human Vision, Visual Processing and Digital Display V*, volume 2179 of *SPIE Proceedings*, pages 117–126, San Jose, CA, February 1994.

[SdSG94] D.G. Sampson, E.A.B da Silva, and M. Ghanbari. Wavelet transform image coding using lattice vector quantisation. *Electronics Letters*, 30(18):1477–1478, September 1994.

[Sha48] Claude E. Shannon. A mathematical theory of communication. *Bell Systems Technical Journal*, 27(3):379–423, July 1948. Continued 27(4):623-656, October 1948.

[Sha59] Claude E. Shannon. Coding theorems for a discrete source with a fidelity criterion. In *IRE National Convention Record*, volume 7, part 4, pages 142–163, New York, NY, March 1959.

[Sig96] Julien Signes. Reducing the codebook size in fractal image compression by geometrical analysis. In Rashid Ansari and Mark J. Smith, editors, *Visual Communications and Image Processing '96*, volume 2727 of *SPIE Proceedings*, pages 1400–1409, 1996.

[Sig97] Julien Signes. Geometrical interpretation of IFS based image coding. *Fractals*, 5 (Supplementary Issue):133–143, July 1997.

[SN96] Gilbert Strang and Truong Nguyen. *Wavelets and Filter Banks*. Wellesley-Cambridge Press, Wellesley, MA, 1996.

[Sol97] Stephen J. Solari. *Digital Video and Audio Compression*. McGraw-Hill, New York City,New York, 1997.

[Sto88] James Storer. *Data Compression*. Computer Science Press, Rockville, MD, 1988.

[Tha96] Nguyen T. Thao. A hybrid fractal-DCT coding scheme for image compression. In *Proceedings ICIP-96 (IEEE International Conference on Image Processing)*, volume I, pages 169–172, Lausanne, Switzerland, September 1996.

[var98] various. *Proceedings of Data Compression Conference 1998*. IEEE Computer Society, Los Alamitos, CA; Piscataway, NJ; Brussels; Tokyo, April 1998.

[Wal91] Gregory K. Wallace. The JPEG still picture compression standard. *Communications of the ACM*, 34(4):30–44, April 1991.

[Wat93] Andrew B. Watson. Visually optimal DCT quantization matrices for individual images. In James A. Storer and Martin Cohn, editors, *Proceedings DCC'93 (IEEE Data Compression Conference)*, pages 178–187, Snowbird, UT, April 1993.

[Way95] Peter Wayner. Strong theoretical steganography. *Cryptologia*, 19(3):285–299, July 1995.

[Way96] Peter Wayner. *Disappearing Cryptography*. AP Professional, Chestnutt Hill, MA, 1996.

[Way97] Peter Wayner. *Digital Copyright Protection*. AP Professional, Chestnutt Hill, MA, 1997.

[WC92] A. Wolfe and A. Chanin. Executing compressed programs on an embedded RISC architecture. In *Proc. Int'l Symp. on Microarchitecture*, 1992.

[WCMP96] X. Wang, E. Chan, M. K. Mandal, and S. Panchanathan. Wavelet-based image coding using nonlinear interpolative vector quantization. *IEEE Transactions on Image Processing*, 5(3):518–522, March 1996.

[WHK93] Kwo-Jyr Wong, Ching-Han Hsu, and C. C. J. Kuo. Fractal-based image coding with polyphase decomposition. In Barry G. Haskell and Hsueh-Ming Hang, editors, *Visual Communications and Image Processing '93*, volume 2094 of *SPIE Proceedings*, pages 1207–1218, 1993.

[WP91] H. R. Wu and F. J. Paoloni. A new two-dimensional fast cosine transform algorithm based on Hou's approach. *IEEE Transactions on Signal Processing*, 39(2):544–546, February 1991.

[ZL77] Jacob Ziv and Abraham Lempel. A universal algorithm for sequential data compression. *IEEE Transactions on Information Theory*, IT-23(3):337–343, May 1977.

[ZY96] Y. Zhao and B. Yuan. A hybrid image compression scheme combining block-based fractal coding and DCT. *Signal Processing: Image Communication*, 8(2):73–78, March 1996.

Index

11172-3, 162

2h1v, 128
2h2v, 128

Abramson, Norman, 23
Adler, Mark, 47
Adobe, 77
Ahmed, N., 106
Aiken, Alex, 72
Allen, James D., 205, 206
arc files, 46
arithmetic coding, 126, 133
arj files, 47
Arps, Ronald B., 198
Atkinson, William D., 215

backward prediction, 146
Bacon, Francis L., 181
Barahav, Zachi, 166
Barnsley, Michael, 165
Barnsley, Michael F., 216–219
baseline, 133
basis, 101
Bell, Timothy, 59
Bertin, Etienne, 167
Betz, Bernard K., 213
bidirectional prediction, 146
Bilgin, Ali, 103
Bird, Peter, 71
Block Layer, 144
book cipher, 172
Bormans, Jan G., 106
Boss, Roger D., 218
Breternitz, Mauricio, 71

Campbell, Joe, 159
Cate, Vincent, 71
Chambers, IV, Lloyd L., 192
Chamzas, Christosdoulos, 182, 202, 203
Chanin, A, 72
Chan, E., 102
Chan, S. C., 106
Chassery, Jean-Marc, 167
Cheng, Bing, 167
Chen, I-Cheng, 71
Chen, Wen-Hsiung, 106
Chevion, Dan S., 202, 204
Chomsky, Noam, 67
Christopoulos, Charilaos A., 106
Chui, Charles, 101
Cleary, John, 59
code compression, 71
Cohen, Albert, 101
Cohn, Martin, 184
Collage Theorem, 166
collisions, 43
Common Intermediate Format (CIF), 142
comp.speech FAQ, 159
complexity, 12

compress, UNIX standard, 47
context-free grammar, 69
context-sensitive grammar, 69
Cornelis, Jan P., 106
Cosman, Pamela C., 94, 103
Cover, Thomas M., 23

da Silva, E.A.B , 102
Daubechies, Ingrid, 101, 112
Davoine, Franck, 167
Demko, Stephen, 165
Devadas, Srinivas, 72
DFT, 2-D, 108
dictionary approaches, introduction, 2
digital signal processors, 101
dilation, 111
dimension reduction, 98
discrete cosine transform, 13, 106, 125, 141
Discrete Fourier Transform, 103
dot product, 107
downsample, 128
DSP, 101
Duttweiler, Donald L., 182, 202, 203

Eastman, Willard L., 184
Elton, John H., 218, 219
entropy, 23
Ernst, Jens, 71
Evans, William, 71
excitation, 161
extended binary trees, 22

Falconer, Ken, 166
Fenwick, Peter, 27
Fiala, Edward R., 187
Finn, Steven G., 180
first-order, 27
Fisher, Yuval, 218
flexible length coding, 73
forward prediction, 146, 148
Fourier, J. B. J., 102
Fralick, S. C., 106
Franz, M., 72
Fraser, Christopher, 71
frequency space, 105
Fukui, Ryo, 205
Furlan, Gilbert, 204

Gailly, Jean-Loup, 47
Gallagher, Robert, 23
George, Glen A., 189, 191, 212
Ghanbari, M., 102
Gibson, Dean K., 190
GIF (Graphics Interchange Format), 94, 126
Goertzel, Gerald, 197, 210
grammar, 67
Graybill, Mark D., 190
Gray, Robert M., 94, 103
Greene, Daniel H., 187
Greibach Normal Form, 176
Gross, Thomas, 71
Group of Blocks Layer, 144
group of pictures, 144
gzip algorithm, 179
gzip files, 47

H.261, 142
Hamming, Richard, 23
Hamzaoui, Raouf, 166
Han, Bin, 102
Hara, Makoto, 210
hash tables, and Lempel-Ziv, 43
Hashiue, Masakazu, 210
Hayes, Monson H., 166
Heckbert, Paul, 95
Henriques, Selwyn, 192

hierarchical encoding, 126
hierarchical transmission, 137
Hirose, Osamu, 211
Hoerning, John S., 183
Hollatz, Scott A., 166
Holtz, Klaus E., 214
Hopcroft, John, 69
Horikawa, Hiroshi, 217
Houde, Donald J., 181
Hou, Hsieh S., 106
Ho, K. L., 106
Huffman coding, 126, 133
Huffman trees, 22
Humphrey, Stanley A., 215
Hwang, J.J., 125, 222
Hürtgen, Bernd, 166

International Standards Organization, 141
International Telephony Union, 141
Interpress, 77
ISO 11172-3, 162
iterated, 166
Iterated Systems, 165
Itoh, Kunihiro, 209
Ivey, Glen E., 189, 191

Jackson, Rory D., 184
Jacobs, Everett W., 218
James, David C., 220
Johannesson, Brian J., 209
Joint Photographic Experts Group, 125
JPEG, 7, 106
JPEG standard, 141
JSteg, 174
Jung, Robert K., 192

Kaiser, Gerald, 101
Kakuse, Katsuharu, 209
Karnin, Ehud D., 166, 198, 202, 204
Katz, Phillip W., 190
Kenner, Hugh, 172
Keutzer, Kurt, 72
Kimura, Tomohiro, 182, 205, 206
Kino, Shigenori, 182, 205, 206
Kistler, T, 72
Kleijn, W. Bastiaan, 158
Knipe, Jason, 102
Kolmogorov Complexity, 12, 23
Kominek, John, 166
Kondo, Yoshie, 209
Kopf, David, 3
Kozuch, Michael, 72

L_1 norm, 146
L_2 norm, 146
Langdon, Jr., Glen G., 49, 180, 194–196, 199, 200, 208, 213, 214
latency, 8
Layer 3, 162
least asymetric filter, 117
Lefurgy, Charles, 71
Lempel, Abraham, 35, 183, 184
Lepsøy, Skjalg, 166
lha files, 47
Liao, Stan, 72
Lincoln, Abraham, 1
Livny, Miron, 124
Li, Xiabo, 102
Loder, Maria A., 207
loop filters, 149
Lopresto, Scott M., 102
lossless, 3
lossless JPEG, 134
lossy, 3
Louis, A.K., 101
Lucco, Steven, 71
luminance quantization matrix, 128
lzh files, 47

Maaß, P., 101
MacCrisken, John E., 186
Macroblock Layer, 144
Malah, David, 166
Mandal, M. K., 102
Mandelbrot, Benoit, 166
Marcellin, Michael W., 103
masking, 160
Mathematica, 101
Mathematica's Wavelet Explorer, 117
Maulsby, D.L., 69
Mazel, David S., 166
McIntosh, Duane E., 180
Meyer, Yves, 101, 112
MIDI files, 103
Miller, Victor S., 187
Mitchell, Joan L., 106, 125, 197–201, 203, 204, 210
Mohiuddin, Kottappuram M. A., 197
Moll, Edward W., 184
Monro, Donald M., 124
Moon, Hyeon J., 102
Moreman, Charles S., 218, 219
motion estimation, 141
Motion-JPEG, 141
MP3, 162
MPEG, 141
MPEG compression standard, 161
MPEG Layer 3, 162
MPEG-1, 142
Mudge, Trevor, 71
Mukherjee, Amar, 181

Nagy, Michael E., 189
Narasimha, Madihally J., 106
Nasrabadi, Nasser M., 102
Natarajan, T., 106
Nelson, Mark, 222
Nevill-Manning, Craig, 67, 69
Nguyen, Truong, 101
Nicolau, Alex, 72
Niv, Ron, 216
Nomizu, Yasuyuki, 207
normalized, 105
Notenboom, Leo A., 188, 191
NTSC, 142
numbers, flexible coding, 73
Nyquist frequency, 159

O'Brien, John T., 188
Øien, Geir E., 166
Olshen, Richard A. , 103
ones compliment, 132
Ono, Fumitaka, 182, 205, 206
Orchard, Michael T., 102
O'Rourke, Hugh, 172
orthogonal, 105
Ozaki, Syoji, 209

PAL, 142
Paliwal, Kuldip K., 158
pallet, 174
Panchanathan, S., 102
Paoloni, F. J., 106
parametric coders, 158
Park, Shang H., 102
Pennebaker, William B., 106, 125, 198–201, 203, 204, 207
Peterson, Allen M., 106
Picture Layer, 144
PkZip, 10
pkzip compression, 47
Pochanayon, Adisak, 42
PostScript, 77
predictive coding, 134
predictive compression, 131
predictive errors, 131
Primiano, Guy A., 218, 219

production, 68
productions, 68
Proebsting, Todd, 71
progressive encoding, 126
progressive transmission, 136
pulse code modulation, 159
pumping lemma, 69

quanta, 87
quantization, 85
quantization, static, 86
Quarter Common Intermediate Format (QCIF), 142

Rackl, Willi K., 184
Ramchandran, Kannan, 102
Ranganathan, N., 192
Rao, K. R., 106, 125, 222
Ratnakar, Viresh, 124
RealAudio, 8
Reese, Craig, 159
regular grammar, 69
regular languages, 69
residual values, 117
Reverse Polish Notation, 78
Rieder, A., 101
RISC code compression, 71
Rissanen, Jorma J., 49, 180, 194–197, 199, 200, 203, 214
run length encoding, 2
Ryan, Robert, 101

Saba, Shinpei, 209
Safranek, Robert, 124
Sampson, D.G., 102
SECAM, 142
selective refinement, 134
self-extracting archive, 7, 12
sequential coding, 126
SEQUITUR, 69
Shannon, Claude, 10, 16, 22
Sherlock, Barry G., 124
Shimoni, Yair, 216
Signes, Julien, 166
significant pel area (SPA), 143
singular value decomposition, 98
singular values, 98
Skodras, Athanasios N., 106
slices, 144
Sloan, Alan D., 216–219
Smith, C. Harrison, 106
Smith, Roger, 71
Snow, Craig A., 186
Solari, Stephen, 125
Source Input Format, 142
Specialists Group on Coding for Visual Telephony, 141
sqz files, 47
static quantization, 86
statistical approaches,introduction, 2
Stiller, Christoph, 166
Storer, James, 40, 211, 221
strange attractors, 167
Strang, Gilbert, 101
Stuffit, 10
subband coding, 160
substitutional compression, 35
Symanski, Thomas, 40
symbols, 70

Tanaka, Kazuyoshi, 210
terminals, 67
Thomas, Kasman E., 193
Thomas, Joy A., 23
Tjiang, Steven, 72
token, 28
Toyokawa, Kazuharu, 200
transition sounds, 161

Travesty, 172
tries, 44
Tsukiyama, Tokuhiro, 209, 211, 219
Two-dimensional Fourier Transform, 108

uc2 files, 47
Ullman, Jeffery, 69
UNIX compress, 47
unvoiced sounds, 161
Upham, Derek, 174
Urabe, Hitoshi, 217

variable length codes, 150
variables, 67
vector, 106
vector quantization, 94
Vetterli, Martin, 94
video compression, 141
VLC, 150
voiced sounds, 161
Voronoi diagram, 97

Walach, Eugeniusz, 202, 204
Wallace, Greg, 106, 125
Wang, Albert, 72
Wang, X., 102
Waterworth, John R., 186
Watson, Andrew B., 124
waveform-approximating solutions, 158
wavelet, 5, 103, 111
Wavelet Explorer, 101, 117
wavelet transform, 101
Wegman, Mark N., 187
Welch, Terry A., 185
Whiting, Douglas L., 189, 191, 212
window function, 111
Witten, Ian, 59, 67, 69
Wolfe, Andrew, 72
Wu, H. R., 106

Yako, Katsutoshi, 217
Yashiki, Hiroshi, 211
YCrCb, 127
Yoshida, Masayuki, 182, 205, 206
YUV, 127

Z files, 47
Zandi, Ahmad, 208
Zhu, Xiaokun, 167
zip files, 47
Ziv, Jakob, 35, 183, 184
zoo files, 47
Zweig, George, 103

Related Titles from Morgan Kaufmann

http://www.mkp.com

- **C++ for Real Programmers**
  Jeff Alger
  1998 ISBN: 0-12-049942-8

- **Tcl/Tk for Real Programmers**
  Clif Flynt
  1999 ISBN: 0-12-261205-1

- **CORBA for Real Programmers**
  Reaz Hoque
  1999 ISBN: 0-12-355590-6